Digital Fashion

Digital fashion has emerged as a multidisciplinary field of knowledge that attracts overlapping interests from academics, researchers, and professionals coming from fashion design, business and technology, computer science, software engineering, animation and gaming, anthropometrics, supply chain management, Industry 4.0, big data, artificial intelligence, and industrial sustainability. This book captures the latest innovations and applications in this field to assist readers in navigating this changing discipline.

- Explores how AI is shaping the fashion industry, featuring a case study on branding.
- Details virtual fashion applications, including virtual try-ons, avatars, 3D image analysis, and digital pattern cutting.
- Examines engineered reality in fashion, including VR and XR.
- Discusses fashion and the metaverse.
- Covers phygitals and smart wearables.

Offering an overview of the cutting-edge in digital fashion, this book explores topics that will appeal to a variety of readers including fashion and textile technologists, scientists, and engineers interested in anthropometry for a variety of disciplines as well as development of VR/AR and AI tools, especially for clothing end uses.

Abu Sadat Muhammad Sayem is Fellow of the Higher Education Academy, England, Fellow and Council Member of the Textile Institute, and Fellow of the Royal Society of Arts. He is an alumnus of the UK Commonwealth Scholarship Commission and the German Academic Exchange Programme, commonly known as DAAD. He obtained a PhD from the University of Manchester, an MSc from the Technische Universität Dresden, and a BSc from the University of Dhaka. He has over 20 years of professional experience of working in different roles in industry and academia, including the role of head of the department. He led the Digital Fashion Network project (2023–2024) funded by the Arts & Humanities Research Council (AHRC) in the UK. He currently works at the Loughborough University. Dr. Sayem is a member of the Textile Institute's publication committee and edits the Textile Institute Professional Publication Book Series published by CRC Press/Taylor & Francis.

Textile Institute Professional Publications

Series Editor: Abu Sadat Muhammad Sayem, *The Textile Institute*

Medical Textles
Holly Morris and Richard Murray

Odour in Textiles
Generation and Control
G. Thilagavathi and R. Rathinamoorthy

Principles of Textile Printing
Asim Kumar Roy Choudhury

Solar Textiles
The Flexible Solution for Solar Power
John Wilson and Robert Mather

Digital Fashion Innovations:
Advances in Design, Simulation, and Industry
Abu Sadat Muhammad Sayem

Compression Textiles for Medical, Sports, and Allied Applications
Nimesh Kankariya and René Rossi

Microfibre Pollution from Textiles
Research Advances and Mitigation Strategies
Edited by R. Rathinamoorthy and S. Raja Balasaraswathi

Advances in Renewable Natural Materials for Textile Sustainability
Edited by Mohammad Shahid, Aranya Mallick, and Sayandeep Debnath

Biotechnology Approaches in Textile Technology
Progress and Trends
Edited by Mohd Yusuf and Mohammad Shahid

Digital Fashion
Artificial Intelligence, Virtual, and Other Emerging Applications
Edited by Abu Sadat Muhammad Sayem

For more information about this series, please visit: www.routledge.com/Textile-Institute-Professional-Publications/book-series/TIPP

Digital Fashion

Artificial Intelligence, Virtual, and Other Emerging Applications

Edited by Abu Sadat Muhammad Sayem

CRC Press
Taylor & Francis Group
Boca Raton London New York

CRC Press is an imprint of the
Taylor & Francis Group, an **informa** business

Designed cover image: © 2026 Shutterstock

First edition published 2026
by CRC Press
2385 NW Executive Center Drive, Suite 320, Boca Raton FL 33431

and by CRC Press
4 Park Square, Milton Park, Abingdon, Oxon, OX14 4RN

CRC Press is an imprint of Taylor & Francis Group, LLC

This work was supported by the Arts and Humanities Research Council [grant number AH/T011483/1].

ISBN: 978-1-032-97616-7 (hbk)
ISBN: 978-1-032-97615-0 (pbk)
ISBN: 978-1-003-59448-2 (ebk)

DOI: 10.1201/9781003594482

For Product Safety Concerns and Information please contact our
EU representative: GPSR@taylorandfrancis.com
Taylor & Francis Verlag GmbH,
Kaufingerstraße 24,
80331 München, Germany.

Contents

Contributor List

Eleanor Bandey
De Montfort University
Leicester, United Kingdom

Fatma Baytar
Cornell University
New York, United States

Andrew Bravin
Sheridans
London, United Kingdom

Zowie Broach
Royal College of Art
London, United Kingdom

Karen Cross
Robert Gordon University
Aberdeen, United Kingdom

Morolake Dairo
Manchester Metropolitan University
Manchester, United Kingdom

Karen Deacon
De Montfort University
Leicester, United Kingdom

Philip Delamore
Royal College of Art
London, United Kingdom

Daniel Drak
The New School
New York, United States

Dawn Ellams
Royal College of Art
London, United Kingdom

Jannatul Fardus
American International University–Bangladesh
Dhaka, Bangladesh

Paul Gough
Footfalls & Heartbeats
Nottingham, United Kingdom

Carolyn Hardaker
De Montfort University
Leicester, United Kingdom

Md. Mahmudul Hasan
Bangladesh University of Textiles
Dhaka, Bangladesh

Maxi Heitmayer
University of the Arts London
London, United Kingdom

Jalaj Hora
Manchester Metropolitan University
Manchester, United Kingdom

Grace Hsieh
Shih Chien University
Taipei, Taiwan

Feng-Yu Hsu
Shih Chien University
Taipei, Taiwan

Emily Huggard
The New School
New York, United States

Alice Janssens
University of Southampton
Southampton, United Kingdom

Yang Jiang
Hull University
Hull, United Kingdom

Naila Sharmin Kanta
Bangladesh University of Textiles
Dhaka, Bangladesh

Fatma Kalaoglu
Istanbul Technical University
Istanbul, Turkey

Jeong Min Kho
Yonsei University
Seoul, South Korea

Si Eun Kim
Changwon National University
Changwon, South Korea

Se Jin Kim
Changwon National University
Changwon, South Korea

Ruobing Li
Yonsei University
Seoul, South Korea

Linda Lightley
Northumbria University
Tyne, United Kingdom

Chenming Lin
Liverpool University
Liverpool, United Kingdom

Lyndsay Mesjar
Robert Gordon University
Aberdeen, United Kingdom

Sharon Nunoo
Manchester Metropolitan University
Manchester, United Kingdom

Evrim Buyukaslan Oosterom
Istanbul Bilgi University
Istanbul, Turkey

Gary Pogue
Northumbria University
Tyne, United Kingdom

Susan Postlethwaite
Manchester Metropolitan University
Manchester, United Kingdom

Sheng-feng Qin
Northumbria University
Tyne, United Kingdom

Md. Mahbubor Rahman
Bangladesh University of Textiles
Dhaka, Bangladesh

Eleanor Rockett
University of the Arts London
London, United Kingdom

Erika Aline Rox
Politecnico di Milano
Milano, Italy

Natalia Särmäkari
Metropolia University
Helsinki, Finland

Abu Sadat Muhammad Sayem
Loughborough University
Loughborough, United Kingdom

Yuyuan Shi
Northumbria University
Tyne, United Kingdom

Lucie Shilton
University of the Arts London
London, United Kingdom

Ghada Soliman
Manchester University
Manchester, United Kingdom

Marta Sron
University of the Arts London
London, United Kingdom

Josie Steed
Robert Gordon University
Aberdeen, United Kingdom

Brian Taylor
University of Alabama
Alabama, United States

Zhe Wang
Manchester Metropolitan University
Manchester, United Kingdom

Guobin Xia
Liverpool University
Liverpool, United Kingdom

Yanna Jessica Xu
Independent Researcher

Eun Kyoung Yang
Yonsei University
Seoul, South Korea

Rong Zheng
Donghua University
Shanghai, China

About the Editor

Trained as a textile technologist at the textile college of the University of Dhaka, Dr. Abu Sadat Muhammad Sayem gained expertise in fashion technology through his master's study at the Technische Universität Dresden and PhD study at the University of Manchester. He is a fellow and trustee of the Textile Institute, fellow of the Higher Education Academy (UK), fellow of the Royal Society of Arts, as well as an alumnus of the Commonwealth Scholarship Commission (UK) and the Deutscher Akademischer Austauschdienst (DAAD, Germany). He led the "Digital Fashion Network" project initiated with funding from Arts and Humanities Research Council (AHRC). He serves the Textile Institute Professional Publication Series as the Editor-in-Chief. He has been chairing the Digital Fashion Innovation symposia and conference in the UK since 2020, the Fashion-Textiles-Wearables congress in the UK since 2021, and the Textile Research Conference in Bangladesh since 2014. He has over 20 years of experience in academic and research leadership at multi-national and multi-continental higher education (HE) settings, including the roles of head of the department, head of faculty research centre, and principal investigator of national and international projects with UKRI funding, high-profile research outputs including edited books, journal special issue and journal articles, and process-product innovations in digital prototyping, wearable technologies, and sustainable manufacturing.

Preface

This book *Digital Fashion: Artificial Intelligence, Virtual, and Other Emerging Applications* discusses the state-of-the-art digital developments that are disrupting the fashion industry and reshaping the way we imagine, design, produce, experience, and perceive fashion. It is structured into five thematic parts, each addressing a critical dimension of this evolving landscape and bringing together interdisciplinary perspectives to reflect the complexity and opportunity found in the digital fashion frontier.

Part A is composed of five chapters (1–5) and covers how AI is remaking design processes, redefining brand strategies, and enhancing inclusivity (Figure 1).

Part B explains virtual fashion applications by delving into the issues like body satisfaction and identity making in the context of virtual bodies, and the simulation tools that are transforming both fashion education and heritage preservation. This part includes four chapters (6–9), as presented in Figure 2.

Part C shifts the focus to immersive and extended realities, examining their implications for branding, sustainability, and consumer experience. The three chapters (10–12) included in this section explore immersive experiences through the lens of consumers and technologies (see Figure 3).

Part D looks into the convergence of fashion, gaming, and Web3, highlighting new models of fashion creativity, community, and monetisation. This part has four

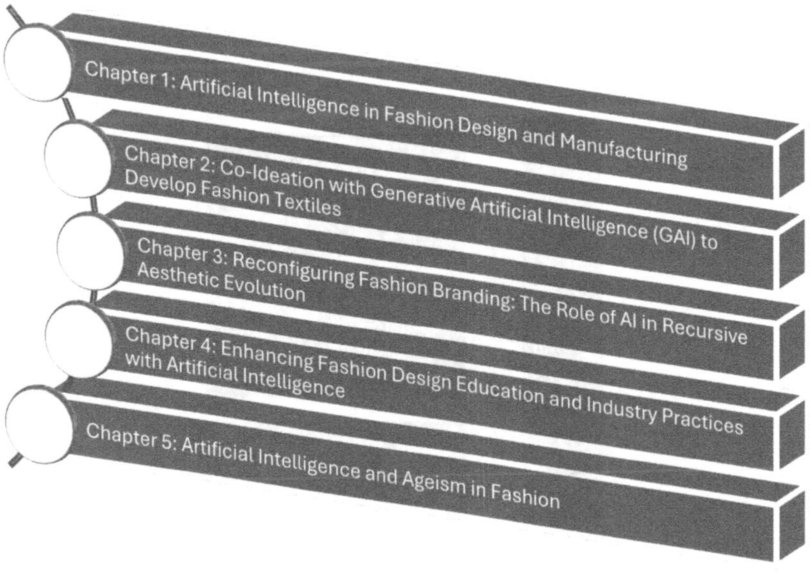

Figure FM.1 Construct of Part A: artificial intelligence shaping the fashion industry.

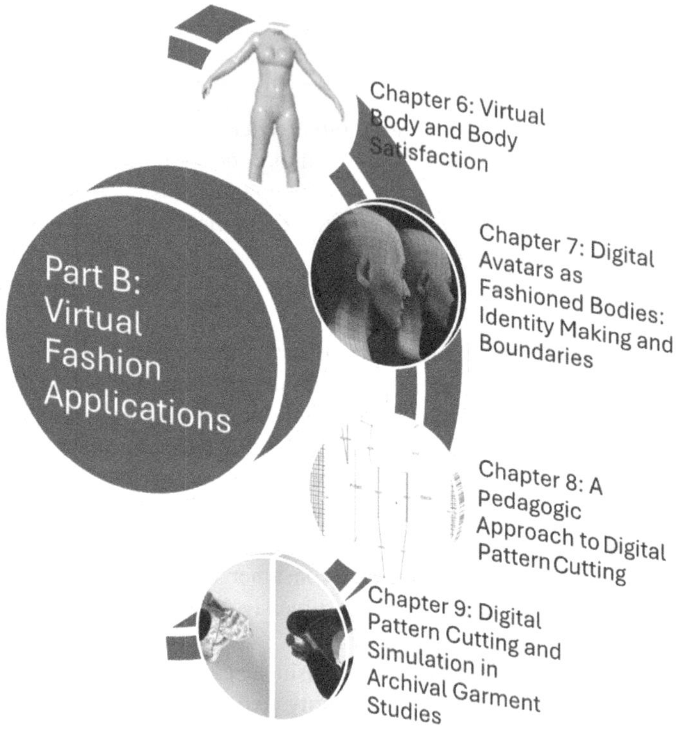

Figure FM.2 Elements of virtual fashion applications.

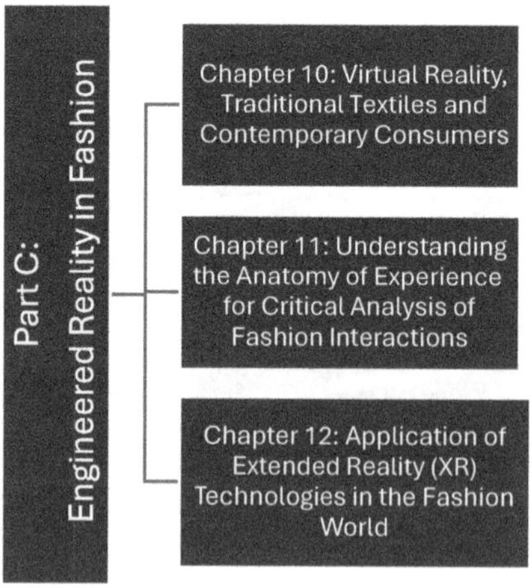

Figure FM.3 The coverage of Part C: engineered reality in fashion.

chapters (13–16) that discuss technology, application, business, and community in the context of fashion in metaverse (see Figure 4).

Part E consists of two chapters (Figure 5). Chapter 17 reflects on interdisciplinary teaching and realisation of electronic garments, the phygital side of the digital fashion. Chapter 18 presents a critical assessment of the digital fashion field's current state and future directions.

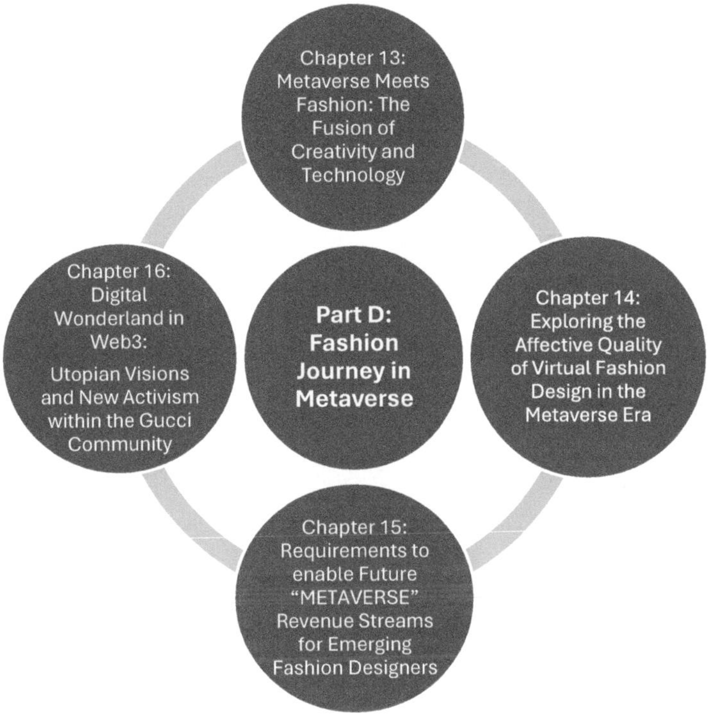

Figure FM.4 Contents of part D: fashion journey in metaverse.

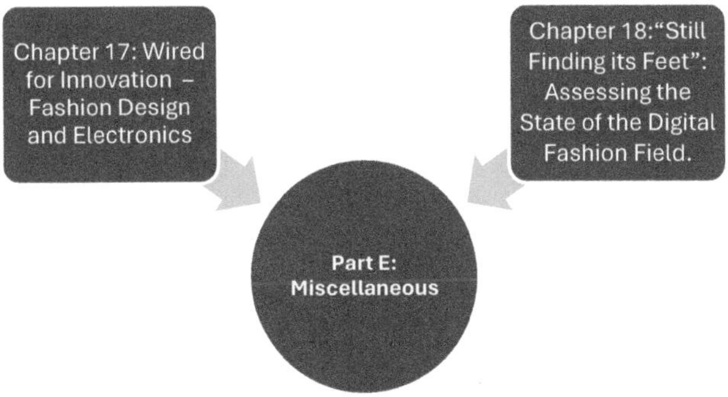

Figure FM.5 Topics of Part E: miscellaneous.

While reading this book, readers will discover a multifaceted view of digital fashion as a dynamic, interdisciplinary, and rapidly evolving domain. The book offers both rich academic inquiry and real-world examples grounded in practical applications. I hope it will serve as a valuable resource for students seeking a foundational understanding, academics pursuing research directions, and professionals adapting to emerging industry standards.

Abu Sadat Muhammad Sayem

Part A

Artificial Intelligence Shaping Fashion Industry

1 Artificial Intelligence in Fashion Design and Manufacturing

Abu Sadat Muhammad Sayem, Susan Postlethwaite, and Jalaj Hora

1.1 INTRODUCTION

The science of artificial intelligence (AI) has made significant advancement over the last seven decades and found remarkable applications in many of our personal and industrial activities. Starting with the advent of Symbolic AI in 1960s, it has gone through the development phases of "Knowledge-based Expert System" in the 1970s, "Data-driven AI" in the 1980s, "AI for Optimisation" in the 1990s, "Deep Learning" in the 2010s and reached to the level of "Cognitive AI" approaching to the capability of human-level intelligence to face uncertainty and self-learning and reliability to be deployed commercially to ease or enhance different human-led activities (Mypati et al., 2023). Notable works have been done in the last two decades to develop AI tools for the fashion industry. Table 1.1 lists the reviews on the development and application of AI in the fashion domain published in the last five years. Most of them targeted technical readers from computer and engineering fields, while only a few of them catered general readers by avoiding or simplifying technical jargons and formulae.

The research activities done in the last few decades to develop AI tools for the fashion industry focus on the development of a) AI-enabled computer-aided design (CAD) systems, b) digital shopping and augmented reality try-on systems, c) fashion e-business tools and d) driving automation in manufacturing, including robotics (Zhang & Liu, 2024; Luce, 2019). This chapter takes the route of a traditional review and covers both research works and commercial developments on AI application in fashion design and manufacturing targeting both general and fashion readers.

1.2 AI TECHNIQUES AND APPLICATIONS

AI seeks to reproduce human intelligence in non-living objects such as machines. But non-living objects cannot experience and learn, so it necessary to feed them with data sources of relevant experiences. This process is known as machine learning (ML). A subcategory of ML is neural network (NN), which was initially modelled based on general perception of neurons' activity in the human brain. When this learning process involves a large neural network, it is known as deep learning (DL).

DOI: 10.1201/9781003594482-2

TABLE 1.1
Existing Reviews on AI Relevant to Fashion

#	Authors (Year)	Journal's Name	Review Type	Search System Used	Topics Covered	Target Readers
1	Kouslis et al. (2024)	*Electronic Commerce Research*	Semi-systematic	Google Scholar	AI research on fashion, AI methods, dataset used, product types covered in research	General readers
2	Guo et al. (2023)	*IEEE Access*	Traditional	Not mentioned	Fashion detection, synthesis and recommendation	Technical readers
3	Chen et al. (2023)	*IEE Signal Processing*	Traditional	Not mentioned	Fashion analysis, recommendation and synthesis	Technical readers
4	Deldjoo et al. (2023)	*ACM Computing Surveys*	Traditional	DBLP computer science bibliography	Fashion recommender systems	Technical readers
5	Landim et al. (2021)	*International Journal of Fashion Design, Technology and Education*	Systematic	Multiple databases	Chatbot for fashion e-commerce	Fashion readers
6	Pillarisetty and Mishra (2022)	*International Journal of E-business Research*	Semi-systematic	Scopus	Customer experience and e-satisfaction	General readers
7	Noor et al. (2021)	*Journal of Textile Institute*	Traditional	Not mentioned	AI application in the apparel industry	Fashion readers
8	Chang et al. (2021)	*ACM Computer Survey*	Traditional	Not mentioned	Computer vision and fashion	Technical readers
9	Chang et al., 2021	*Lecture Notes in Networks and Systems*	Systematic	Multiple databases	Machine learning for fashion trend forecasting	Technical readers
10	Lomas et al. (2021)	*6th International Conference on Smart and Sustainable Technologies*	Systematic	Web of science	AI in fashion retail	Technical readers
11	Chakraborty et al. (2021)	*Informatics*	Traditional	Not mentioned	Fashion recommendation system	Technical readers
12	Gu et al. (2020)	*Information Processing and Management*	Traditional	Not mentioned	Fashion analysis	General readers
13	Giri et al. (2019)	*IEEE Access*	Systematic	Scopus and web of science	AI applications in the fashion and apparel industry	General readers
14	Zhang et al. (2019)	*IEEE*	Traditional	Not mentioned	AI in fashion landmark detection	Technical readers
15	Yang et al. (2007)	International Journal of Clothing Science and Technology	Traditional	Not mentioned	Pattern cutting	Fashion readers

In other words, this is an AI technique for a machine to gain a certain level of intelligence. In literature, they are also referred to as AI methods or AI models.

The prevailing AI techniques can be categorised as a) symbolic AI and b) computational AI (Guo & Wong, 2013). The symbolic AI techniques are based on development of knowledge-based systems that include expert system (ES) and case-based reasoning system (RS). Whereas computational AI techniques deal with the development of nature-inspired computation approaches, such as evolutionary computations (ECs), artificial neural networks (ANNs) and fuzzy logic (FL) systems. Mypati et al. (2023) identified ten classes of AI techniques relevant to general manufacturing industry. They are a) search algorithms, b) knowledge-based expert systems (KBESs), c) shallow ML algorithms, d) FL, e) regression algorithms, f) artificial neural networks (ANNs), g) unsupervised algorithms, h) optimisation algorithms, i) DL algorithms and j) reinforcement learning (RL). Giri et al. (2019) identified five broad classes of AI techniques applied for researching AI applications in fashion, which are a) machine learning (ML), b) decision support system (DSS), c) expert system (ES), d) optimisation techniques and e) computer vision (CV) or machine vision. Luce (2019) reported the neural networks (NNs), generative adversarial networks (GANs) and data mining (DM) as the important AI techniques for industrial applications. The AI-assisted industrial activities or systems can be generally grouped as a) natural language processing (NLP), b) computer vision (CV), c) predictive analytics and d) automated system and robotics. Figure 1.1 presents different application areas of AI relevant to the fashion industry.

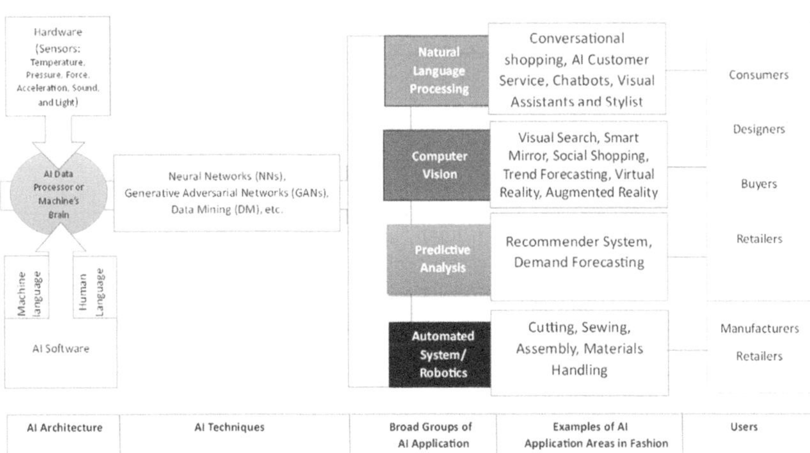

FIGURE 1.1 AI architecture and examples of AI application areas in fashion.

1.3 AI FOR FASHION DESIGN AND DEVELOPMENT

Fashion design is a controlled and guided creative process (see Figure 1.2). Unlike fine artists, the creativity of fashion designers is influenced by several external factors, such as a) trend information (e.g. trend analysis, popularity prediction, colour forecasting), b) customer demography (e.g. size, shape, gender and age), c) material information (e.g. trends, price and sustainability) and market information (e,g. segment, sales analysis and forecasting). These factors influence the design thinking process within designers' brains and guide the designers to develop fashion concepts and ultimate final designs. AI has been suggested and demonstrated in both fashion concept generation and design development processes. In particular, AI-assisted image generation tools are being developed and promoted to automate the fashion design process for making designers free from the time-consuming sketching, drawing and photo-editing tasks.

The advances in the generative artificial intelligence (GAI) through ML frameworks like GANs and variational autoencoders (VAEs) have opened the windows for AI-driven high-quality image generation. The AI-driven CAD systems incorporated with GAN-based image generators can facilitate image-to-image translation by utilising an image database as input to produce new images, taking ideas from source images but still looking different.

GAN-based AI-image generators have been attempted to support the fashion design process by assisting fashion sketches and mapping 2D garments images on human figures (Zhang & Liu, 2024; Liu et al., 2019; Luce, 2019). The recently developed "diffusion modelling (DM)" based text-to-image GAI tools are capable

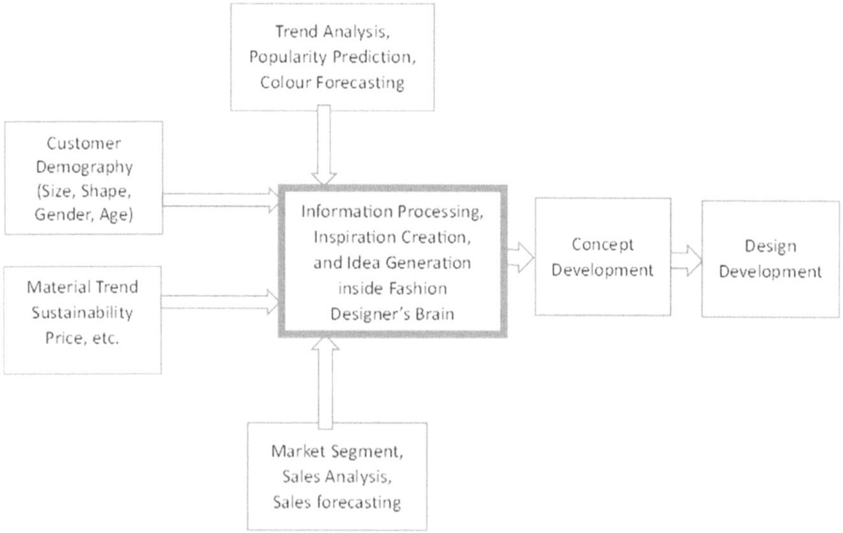

FIGURE 1.2 Fashion design as a controlled and guided creative process.

TABLE 1.2
GAI CAD System vs Fashion CAD Systems Available on the Market

		Input Prompt			Output Design Format				
#	CAD Systems	Text to Image	Image to Image	Text + Image to Image	2D Fashion Design	2D Pattern Design	3D Image	3D Modelling	GAI Model "Stable Diffusion"
1	Midjourney	✓		✓					✓
2	DALL-E	✓							✓
3	CALA	✓	✓	✓					✓
4	AiDA3.0	✓	✓	✓	✓				✓
5	Clo3D				✓	✓	✓	✓	
6	Style3D				✓	✓	✓	✓	
7	Vsticher				✓	✓	✓	✓	
8	Modaris				✓	✓	✓	✓	

of generating high resolution images in high speed in response to human natural language prompts (Borji, 2022). Among the AI-assisted online CAD systems, that are built on DM-based GAI tools, the Midjourney, DALL-E, CALA and AiDA3.0 have drawn notable attention from the field of art and design. Table 1.2 compares the features like input prompts and output design formats of these CAD systems against traditional 3D CAD systems – Clo3D, Style3D, Vsticher and Modaris – used in the fashion industry.

Almost of all research works done on AI implementation in fashion and published in the last two decades focused on development of AI algorithms for fashion application and testing those. Only few works discussed and demonstrated the application of readily available AI CAD systems for fashion design. The following two subsections highlight the software development works and application research works done in this domain.

1.4 SOFTWARE DEVELOPMENT RESEARCH

The development of AI-enabled design software tools has been contributed by the research activities done in the following areas – identification of fashion elements by AI and synthesis of fashion image and design (Guo et al., 2023). Table 1.3 provides a list of the sub-themes associated with these broad themes and a non-comprehensive list of examples of relevant AI techniques that have been developed, proposed and demonstrated in the recent years.

1.4.1 Identification of Fashion Elements

There are trillions of images of different objects available on the internet or in different digital databases. But how would a software system identify which one is an image of a fashion item (e.g. shirt, trousers, socks, etc.) or which one is an image of a

TABLE 1.3
Research Themes Contributing to the Developments of AI-Enabled Design Software Systems and Examples of Relevant AI Techniques.

Research Themes	Sub-Themes	AI Techniques Developed and Applied
Identification of fashion element	Landmark detection	Attentive fashion grammar network (Wang et al., 2018), Part affinity fields (PAFs) (Huang et al., 2019); Multi-depth dilated (MDD) network (Yu & Koltun, 2015), Online hard key points mining (OHKM) (Chen et al., 2018) FashionNet (Liu et al., 2016a) Deep Fashion Alignment (DFA) (Liu et al., 2016b) Fashion Pose machine (FPM) (Hu et al., 2018), Landmark heatmaps (LH) (Liu & Lu, 2019), Spatial-aware Non-local block (SANLb)(Tang et al., 2019), Layout Graph reasoning model (Yu et al., 2019), Bidirectional Convolutional Recurrent Neural Network ((BCRNN) (Wang et al., 2018), DeepLandmark Network (DLAN) (Yan et al., 2017)
	Fashion parsing	Contextualised Convolutional Neural Network (Co-CNN) (Liang et al., 2015), Mask R-CNN (Ruan et al., 2019), Matching CNN (M-CNN) (Liu et al., 2015) Clothes co-parsing (CCP) (Liang et al., 2016)
	Item retrieval	FashionVLP (Goenka et al., 2022), Dual Attribute-aware Ranking Network (DARN) (Huang et al., 2015), Fashion SearchNet model (Ak et al., 2018)
Synthesis of fashion design	Image-based synthesis	End-to-end feed-forward neural network (Jiang & Fu, 2017), FashionG Framework (Jiang & Fu, 2021), Multi-modal unsupervised image-to-image translation (MUNIT) (Huang et al., 2018), Texture and shape disentangled GAN (TSD-GAN) (Yan et al., 2023c) Heatmap guided semantic disentanglement GAN (HSD-GAN) (Yan et al., 2023b) FadGAN based on VAEs conditional GANs (Yan et al., 2023d), Diffusion model DiffFashion (Cao et al., 2024)
	Sketch-based synthesis	TextureGAN (Xian et al., 2018), FashionGAN (Cui et al., 2018), Fashion editing GAN (FE-GAN) (Dong et al., 2020), Conditional feature interaction (CFI) (Yan et al., 2023a)
	Text-based synthesis	Feature-wise linear modulation (FiLM) (Günel et al., 2018), Text and image to image (CTI-IR) (Zhang et al., 2022)
	Multimodal based synthesis	Non-auto regressive generation (NAR) (Zhang et al., 2021)

car or any other object? The very first step in the development of AI-enabled design tools or image analysis tools to train the software systems with logical clues so that they can identify a fashion object out of many images. The AI algorithms developed and proposed for the identification of fashion elements can be classified into three subgroups – landmark detection, fashion parsing and item retrieval. These algorithms train software to learn about different fashion-related attributes, elements and categories within images. Thus, it can analyse images by itself to identify clothing items, styles and patterns, etc. and become a great assistant to designers by removing the time-consuming analysis process and granting them more time for creative aspects of fashion design such as ideation and concept development.

1.4.1.1 Landmark Detection

The landmark detection algorithms can identify landmarks on human body such as waist, hip shoulder points, etc. and landmarks on clothing items such as cuff, hemline, neckband, etc. (Zhang et al., 2019). Figure 1.3 illustrates the AI landmark detection on a t-shirt based on an image input. DL techniques found successful application in landmark detection research. Zhang et al. (2019) classified the AI algorithms developed for landmark detections as regression-based models, pose estimation models, attentive models and constraints-based models (see Table 1.4). Other notable models that are not included in Table 1.4 are Attentive Fashion Grammar Network (Wang et al., 2018), part affinity fields (PAFs) (Huang et al. (2019), multi-depth dilated (MDD) network (Yu & Koltun, 2015) and online hard key points mining (OHKM) (Chen et al., 2018).

1.4.1.2 Fashion Parsing

Fashion parsing, also known as clothing/human parsing, is the process of identifying the correlation between garments and human bodies to facilitate garment mapping on any specific body. This focuses on pixel-level computation to identify detailed features of a fashion image, such as hand, head, dress and pants, etc. (Gu et al., 2020). Figure 1.4 illustrates a fashion parsing process where the shirts and trousers of an input image is changed with differently coloured garments. Notable AI models demonstrated for fashion parsing are Contextualised Convolutional Neural Network (Co-CNN) (Liang et al., 2015), Mask R-CNN (Ruan et al., 2019), Matching CNN (M-CNN) (Liu et al., 2015) and Clothes co-parsing (CCP) (Liang et al., 2016). Gu et al. (2020) classified the contemporary fashion parsing models as graphical models, non-parametric models, parselets representation models, CNN models and adversarial models.

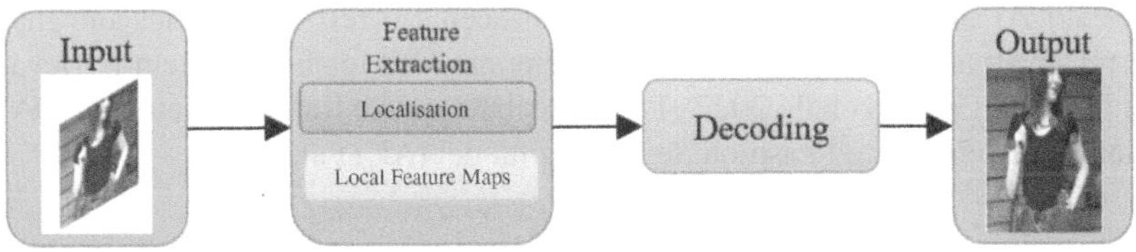

FIGURE 1.3 Example of landmark detention process (red dots in right box indicate identified landmarks on a t-shirt).

(Source: Guo et al., 2023, reproduced under Creative Common Licence)

TABLE 1.4
Group and Examples of AI Algorithms for Landmark Detections

Landmark Detection AI Model Group	Examples
Regression-based models	FashionNet (Liu et al., 2016a) Deep Fashion Alignment (DFA) (Liu et al., 2016b)
Pose estimation models	Fashion Pose machine (FPM) (Hu et al., 2018),
Attentive models	Landmark heatmaps (LH) (Liu & Lu, 2019), Spatial-aware Non-local block (SANLb) (Tang et al., 2019),
Constraints-based models	Layout Graph reasoning model (Yu et al., 2019), Bidirectional Convolutional Recurrent Neural Network (BCRNN) (Wang et al., 2018), DeepLandmark Network (DLAN) (Yan et al., 2017)

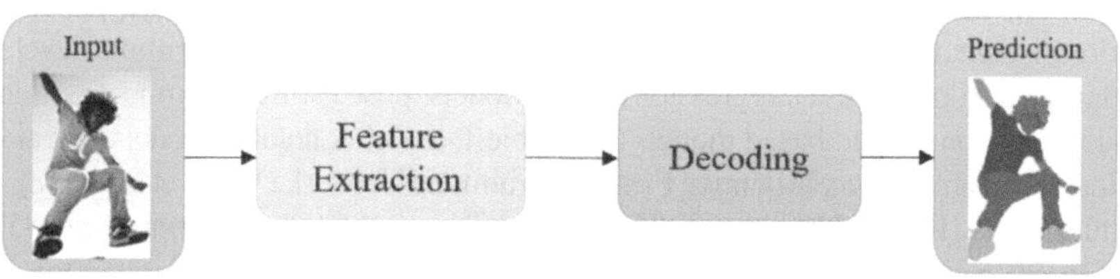

FIGURE 1.4 Example of fashion parsing process.

(Source: Guo et al., 2023, reproduced under Creative Common Licence)

1.4.1.3 Item Retrieval

After a software system is capable of landmark detection and cloth/human parsing, the next important feature the system needs is the item retrieval ability to identify a right item from a source database. Item retrieval is the process of image search and finding a right garment from a source data set and separating it from others. An example of item retrieval process is presented by Kiapour et al. (2015). They developed an algorithm called "Exact Street to Shop" which could identify a fashion item of a given image prompt from a given database of images. Figure 1.5 shows that their system could identify the bag shown in a given picture from a dataset of bags. It has applications both in fashion design and recommendation system. Gu et al. (2020) classified AI item retrieval models as cross-scenario retrieval model and interactive retrieval model. Notable AI models demonstrated for fashion item retrieval are FashionVLP (Goenka et al., 2022), Dual Attribute-aware Ranking Network (DARN) (Huang et al., 2015) and Fashion SearchNet model (Ak et al., 2018).

1.4.2 Synthesis of Fashion Design

AI fashion synthesisers can develop new fashion image or design based on given inputs or prompts. Table 1.5 provides the classification of AI techniques demonstrated in the last decades, and Table 1.3 has presented few examples of those specific

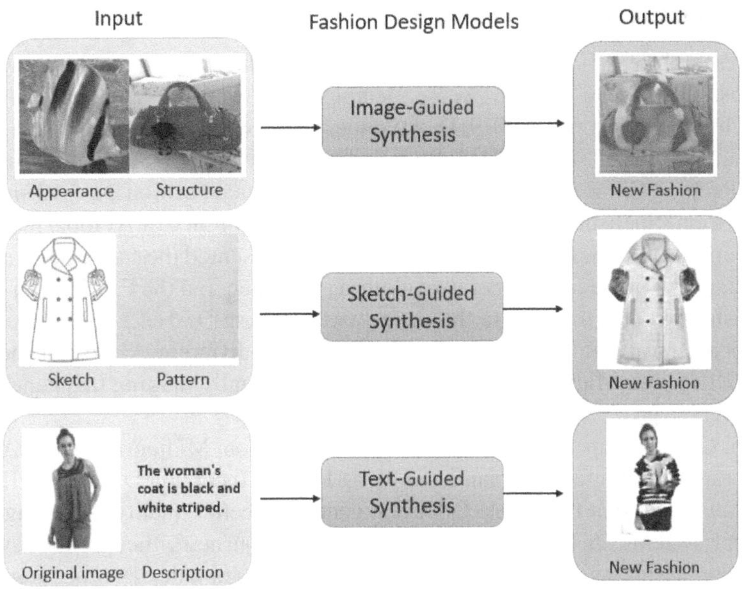

FIGURE 1.5 Example of fashion synthesis with three different types of inputs.

(Source: Guo et al., 2023, reproduced under Creative Common Licence)

TABLE 1.5
Classification of AI Techniques for Fashion Synthesis

Source of Classifications	AI 1.Techniques/Models
Guo et al., 2023	Image-based synthesis model
	Sketch-based synthesis model
	Text-based synthesis model
	Multimodal-based synthesis model
Chang et al., 2021	Style Transfer model
	Pose transformation model
	Physical simulation model
Gu et al., 2020	Pose-guided generative model
	Text guided generative model
	Virtual try-on model
	Image transformation model
	Style generation model

developments. Successful AI models have been presented to generate fashion images based on given prompts in different formats such text, sketch and image. Figure 1.5 exemplifies how AI fashion synthesisers make use of different types of inputs or prompts to generate fashion images, which can work as valuable guidance and source of inspiration for fashion designers.

1.5 APPLICATION RESEARCH

Most research of AI-assisted fashion design is conducted in the field of computer sciences. Only very limited works demonstrated real-life application-oriented research (see Table 1.6). In 2019, DeepVogue, a DL-based AI design system created by a Shanghai-based start-up DeepBlue Technology, won the overall runner-up prize and a "People's Choice Award" at the China International Fashion Design Innovation Competition (Shenshen, 2019). This was a major appearance of AI tools for fashion design application. Särmäkari and Vänskä (2022) presented the case studies of Matti Liimatianen, a pioneering generative clothing designer, and the Fabricant, a digital-only fashion house, to propose the concept of "Cyborg Designer 4.0" representing a physical and digital craftsperson. The advent of different AI design tools like Stable Diffusion, Midjourney and DALL-E2 is gradually making this concept into a reality.

Borji (2022) compared the ability of Stable Diffusion, Midjourney and DALL-E2 to generate photorealistic human faces using Fréchet Inception Distance (FID) metrics. They found that the Stable Diffusion generated better quality face images than the other systems. A DM-based GAI system "Midjourney", though not developed for fashion designers, has been found supportive in fashion design process by accelerating concept and design development to some extent but still relies on designers' expertise and is not to be very helpful for novice and student designers (Zhang & Liu, 2024). Wu and Li (2024) trained a GAN-based GAI with images of knitted textiles, which could generate new images of knitted fabric designs with distinguishable attributes such as stitch type, gauge and appearances and could be used as guiding swatches to knit physical fabric samples by knitters.

Choi et al. (2023) presented a conceptual model of an "AI–human collaborative garment development system". They tested their system by training the StyleGAN2 model with a dataset of 52,000 images collected from 168 leading fashion brands and by generating new fashion images using both image and text-based input prompts. They found that their trained model could generate fashion image with variations in garment silhouette such as full length, sleeve length and neckline and in style such as colour, pattern and print (see Figure 1.6).

TABLE 1.6
Research on Demonstration of AI Application in Fashion Design

#	Reference	AI Tools Applied	Findings/Outcome
1	Zhang and Liu (2024)	Midjourney	AI can assist in fashion design but is not free from limitations
2	Wu and Li (2024)	StyleGAN and RunwayML	AI can generate knitted fabric design good enough to be used reproduced by knitters
3	Borji (2022)	Stable Diffusion, Midjourney and DALL-E2	Stable Diffusion generated better quality face images
4	Choi et al. (2023)	StyleGAN2	Suggested and tested an AI-aided design process

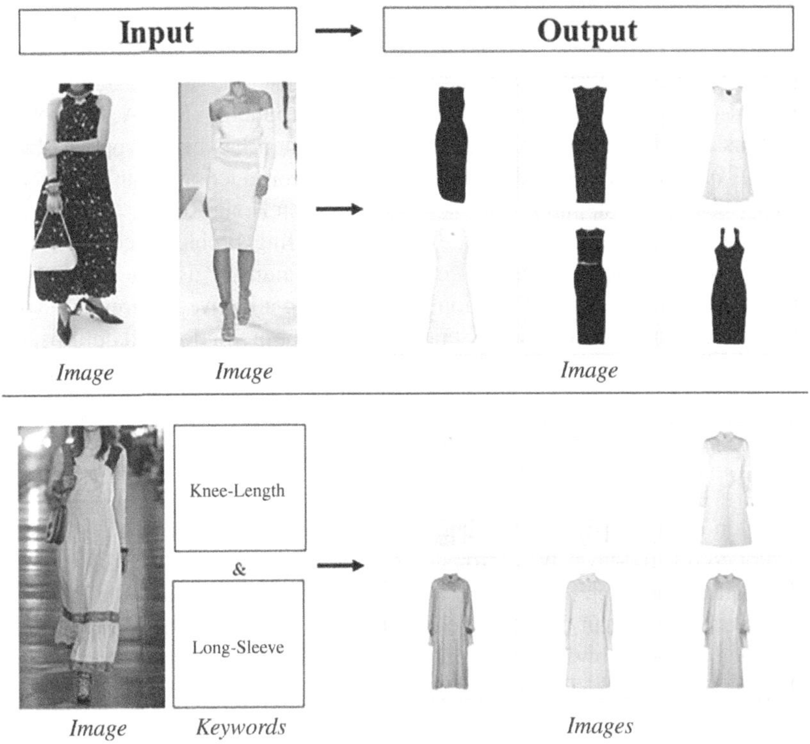

FIGURE 1.6 Examples of Input Prompts and Output Designs using StyleGan2 model.

(Source: Choi et al., 2023, reproduced under Creative Common Licence)

1.6 AI IN FASHION CLOTHING MANUFACTURING

Over the last decades, notable works have been done for applying AI tools into apparel production planning, scheduling, cut order planning, pattern cutting and marker making, which are highlighted in the following subsections.

1.7 PRODUCTION PLANNING, SCHEDULING AND MANAGEMENT

Back in 2001, Wong and Chan (2001) programmed a Genetic Algorithm (GA) based on just-in-time (JIT) philosophy to solve the problem of earliness and tardiness in production scheduling and planning (ETPSP). Their model was found to be effective in minimising inventory and transport cost resulting from earliness and tardiness of order completion. Since then, the GA model gained popularity among the AI researchers for addressing production scheduling problems of the apparel industry. After that significant works have been done in the decade of the 2010s. Guo et al. (2013) developed GAs for production scheduling for maximising resource utilisation and shortening production lead time. Wong et al. (2013a) applied unsupervised and supervised ANN to make decision on the location of apparel manufacturing plants for foreign direct investment and found supervised ANN to be more effective in this purpose. Wong et al. (2013b) applied GAs to optimise operative assignment in the apparel assembly line to achieve a more balanced production line. A set of planning algorithms based on GAs for automatic job allocation has been proposed by Mok et al. (2013), and they claimed that such intelligent production planning algorithms found applications in apparel manufacturing plants in Hong Kong.

Du et al. (2018) investigated the order scheduling problems considering pre-production events and uncertainties in the fashion industry using a multi-objective evolutionary algorithm called nondominated sorting adaptive differential evolution (NSJADE). An intelligent decision support system for prediction and optimisation of production planning was proposed by Ribeiro et al. (2021) with the help of an Evolutionary Multi-objective Optimisation (EMO) algorithm (NSGA-II).

1.8 DIGITAL PATTERN MAKING

Application of AI in pattern drafting goes back to the work of Chan et al. (2003), who applied a three-layer back propagation artificial neural network (BP-ANN) to predict pattern parameters from 3D body measurements. Although the ANN model did not work better than the multiple linear regression model that time, it opened a new branch of research which has continued until today. Ng (2004), as cited in Yang et al. (2007), attempted to simulate the pattern master's expertise using ANN and FL to predict pattern ease. Chen et al. (2009) applied FL ordered weighted averaging (OWA) operator to optimise the estimation of ease allowance of garments. The BP-ANN model that Liu et al. (2017) developed and trained with the anthropometric measurements of 120 young women of the northeastern region of China could predict patterns by utilising body-dimension more accurately than a linear regression model. Very recently, Huang et al. (2024) compared BP-ANN against the linear regression (LR) model in predicting body measurement data for garment pattern making. They trained their BP-ANN model body data of 196 Chinese women from Sichuan Province and fed the same data into their LR model. Similar to the findings from Liu et al. (2017), they also found BP-ANN to be more accurate in predicting body measurements for use in garment pattern making.

Wang et al. (2021) demonstrated a garment pattern design knowledge-base (GPDKB) using support vector regression (SVR), a robust ML method, to support designers by predicting relevant pattern adaptation parameters during pattern drafting. Ma and Yan (2022) proposed a framework of intelligent garment pattern generation system based on BP-ANN. Their proposed system was capable of self-learning, self-organising and designing clothing pattern autonomously. Similar to this system, Wang et al. (2023) demonstrated an interactive pants pattern recommendation system 2022 (IPPRS 2022) built on MATLAB computer language using a 3D digital simulation technique and a series of ML techniques such as radial basis function artificial neural network (RBF ANN), genetic algorithms (GA), probabilistic neural network (PNN) and support vector regression (SVR). Their IPPRS 2022 system can generate patterns directly from 3D garment simulation data such as wearer's body measurement and garment fit preference (i.e. tightness or looseness). They automated the trouser pattern generation process, and the software system they presented could collect a wearer's garment preferences through four interfaces, namely a) anthropometric data acquisition window, b) style selection window, c) ease customisation window, and d) trouser length adjustment window to output customised pattern pieces directly for industrial use (see Figure 1.7).

At a similar time, Oh and Kim (2023) presented an automatic pattern generation system that made use of wearers' body measurements to customise digital pattern pieces and could check the virtual fit of the pattern pieces on representative virtual avatars.

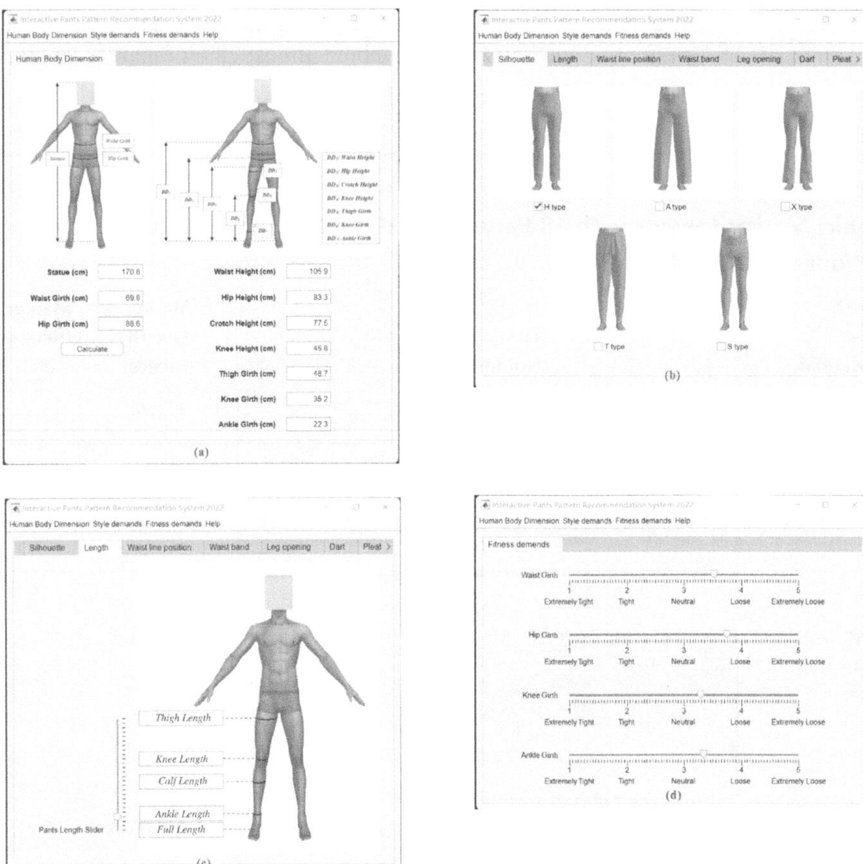

FIGURE 1.7 Interfaces of interactive pants pattern recommendation system 2022 (IPPRS 2022) presented by Wang et al. (2023): a) Anthropometric data acquisition window, b) style selection window, c) ease customisation window, and d) trouser length adjustment window.

(Image reproduced under Creative Common Licence)

1.9 MARKER MAKING AND PLANNING

Marker making deals with the packing problem of irregular shapes in a given dimension of fabric. Wong et al. (2013c) applied the AI methods evolutionary strategies (ES) and neural networks (NN) to optimise marker planning in apparel production. They combined ES with heuristic placement (HP) approach, named as ES + HP approach, and compared its efficiency against GA + HP approach and industry approach (IP) by experimenting with five different marker layouts. Their ES + HP approach was found to be the best in terms of reducing marker length and increasing marker efficiency. Table 1.7 shows one of the five examples of marker experiment in different approaches they applied.

Tsao et al. (2021) ran experiments with different AI algorithms such as GA, simulated annealing (SA) and hybrid genetic algorithm-simulated annealing (HGASA) to shorten the market length of garments to increase marker efficiency. They found that the HGASA performed the best by saving about almost 28% of marker length.

TABLE 1.7
Shirt Marker Layouts with 48 Pattern Pieces in Different Approaches by Wong et al. (2013c)

Method	Marker Layout (Not in Scale)	Market Length (Inches)	Marker Efficiency (%)
Industry Practice	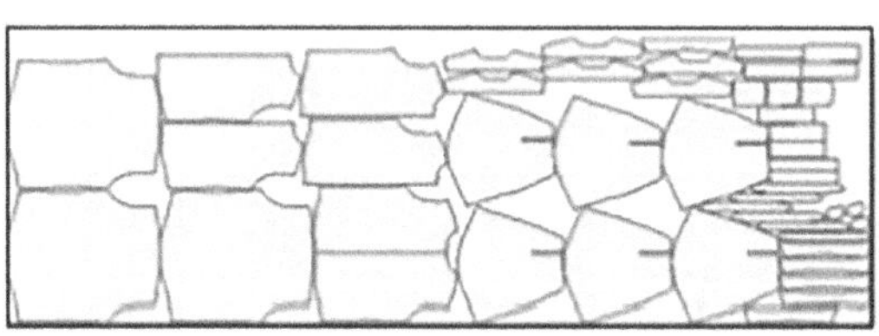	151.78	73.21
AI Method GA + HP	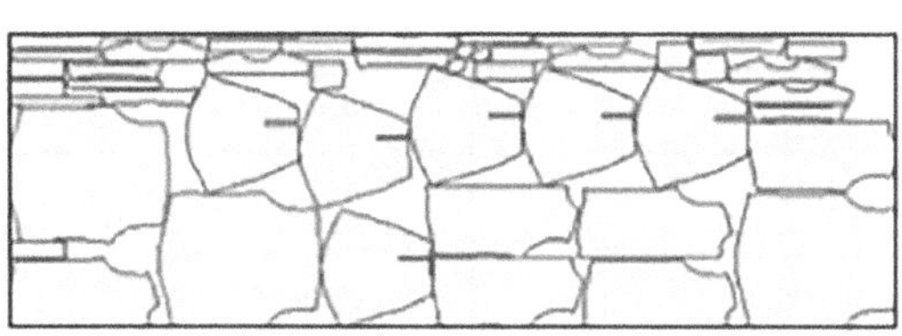	146.94	75.03
AI Method ES + HP	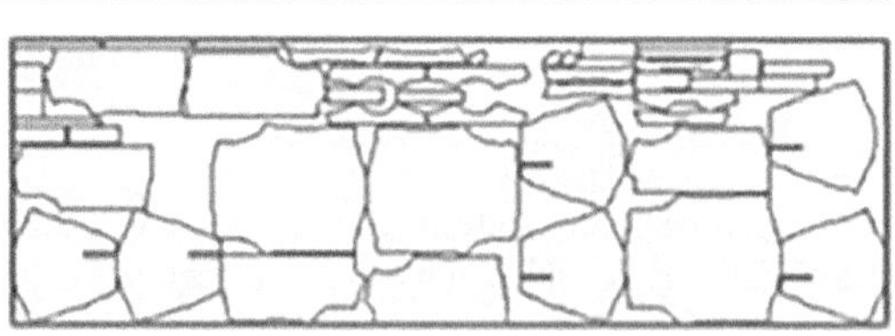	146.60	75.91

N.B: Figures used in table are reproduced from source with permission.

1.10 OPTIMISING FABRIC SPREADING AND CUT ORDER PLANNING (COP)

Patrick et al. (2000) employed GA to formulate and solve roll planning problem in fabric spreading. Their experiment showed GA-assisted optimum roll planning could save about 0.4% of fabrics in the cutting room. Wong et al. (2005) optimised manual fabric spreading and cutting process using GAs. Their GA-based approach was found to be effective in increasing the cut-piece fulfilment rate that could improve the downstream apparel assembly process and in-time delivery of apparel. Later they (Wong et al., 2006) further improved their approach by incorporating the fuzzy set theory with GA to generate the JIT fabric cutting schedule. Mok et al. (2007) carried over the work of Wong et al. (2005, 2006) to develop a genetic optimisation procedure to search fault-tolerant fabric-cutting schedules by minimising makespan and scheduling uncertainties. Bouziri and M'hallah (2007) proposed a new hybrid heuristic to address the difficult and frequently occurring problem in cut order planning in the apparel industry.

Wong and Leung (2008, 2013) addressed the problems of COP in the apparel industry and developed the genetic optimised decision-making model using adaptive evolutionary strategies including GA for reducing both the material costs and the production of additional garments while satisfying time constraints. Wong et al. (2013d) re-explained the application of GA and fuzzy set theory to generate JIT fabric cutting schedules to improve production process and reduce production cost. Tsao et al. (2020, 2024) applied several AI techniques such as simulated annealing-based genetic algorithm (SA-GA), tabu-search-based genetic algorithm (TS-GA), divide-and-search-based Lite heuristic and one-by-one (ObO) heuristic algorithms to solve the COP problems. They concluded that these AI techniques had great potentials for large-scale problem solving in the apparel industry. GA deployed by Alsamarah et al. (2021) was found to be effective in COP and could reduce fabric waste by 2.62% and the number of fabrics spreading layers by over 50%.

The AI application in optimising fabric spreading and cutting has been found effective and reliable over time. Therefore, the AI-enabled software systems for cutting room, for example ***Intellocut*** for Coats Digital, are now commercially available on the market.

1.11 ROBOTIC APPLICATIONS – AUTOMATED SEWING AND FABRIC MANIPULATION

The articulated robots are a common type of industrial robots that are equipped with a manipulator arm having broad degrees of freedom (Luce, 2019). Other notable components of such robots are control system, sensors, actuators, power supply and end effectors. The end effectors play the key role of material handling by picking and moving materials as programmed. For handling soft materials like textile grippers or vacuum cups type, end effectors have found applications in robotic sewing system. The Sewbots from Softwear Automation (Atlanta, USA) use vacuum cup types of end effectors to pick and feed shirt panels under traditional sewing heads for making buttonholes (WSJ, 2015). Whereas the "Sewts.Jupiter" system from Sewts makes use of gripper type end effectors to pick and place fabric pieces on conveyor belt system (Sewts.com). AI is now finding application in these systems.

Koustoumpardis and Aspragathos (2011) combined an Adept Cobra s800 robotic manipulator with a conventional sewing machine to make a robotised sewing system capable of producing straight line lock stitch on rectangular fabric pieces. Their robot arm included a pneumatic gripper controlled by force neural network controller to pull fabrics in a 180-degree direction. Stylios (2013) presented an intelligent sewing system that included a robotic fabric measurement system to measure tensile, buckling, shear, compression and surface properties of fabrics without any operator intervention, a seam laser system to monitor and measure stitching quality during seam operation, a sewing damage system to measure material damage during sewing process and sewing machines equipped with sensors and actuators to optimise production efficiency and quality with minimum operator intervention. They made use of the AI technique, the feedback closed loop of the Neuro-Fuzzy model, which they put into large scale trail at a manufacturing plant in Castleford, UK.

The automated sewing system presented by Ku et al. (2023) made use of computer vision, DL model and open-source Robot Operating System (ROS) to detect the seam line and generate corresponding top stitch path. Their system included a computer vision system with a camera and light sources and a custom-built sewing machine. They used an external PC to control their sewing system which had a translation speed of 60 mm/s. They proposed the use of a collaborative robot with a gripper end effector to automate material handling and transport.

The commercial sewing machine Jack A5E-A from JackSew (Jack Tech, 2023) offers an Adaptive Fabric System (AFS) built with an Octopus Nine-brain AI Chip having 10,000 times/second nine-brain computing power and trained with 200,000 sets of data models. It can sew heavy, light and elastic fabrics automatically without needing any manual adjustments.

Matas et al. (2018) demonstrated the pioneering work of applying deep reinforced learning (DRL) algorithm to train a robot arm (Kinova Mico 7DOF robotic arm) to perform three long horizon tasks such as a) folding a towel to a tape mark, b) diagonal folding of a face towel and c) draping a small towel over a hanger.

Fu et al. (2021) applied the Composite Reward Deep Deterministic Policy Gradient (CR-DDPG) method for developing a process learning framework for a robot to perform fabric manipulation. DL algorithm for dual arm robotic fabric grasping and manipulation was also demonstrated by Gu et al. (2023). Such features of robot training for fabric manipulation are gradually being adopted in systems for commercial applications. AI sewing systems from SewingDX, a spin-off of the Centre for Transformative Garment Production (TransGP) of the University of Hong Kong and Tohoku University, make use of a robotic arm, AccuPick gripper and vision system to perform fixture free, pucker free automatic fabric placement and sewing.

1.12 CONCLUSION

This chapter has provided a simplified overview of the research and development works that have demonstrated the successful applications of AI in fashion design and garment manufacturing. Different AI tools providing solutions to the fashion industry are appearing on the market every week. It is often difficult for the users and fashion professionals to judge and identify the right AI tools for me. Therefore, AI literacy has become an essential skill set for the fashion professionals. Fashion courses are adopting AI tools into their curricula gradually. This chapter will help educators, students and professionals to understand the fundamental AI applications in fashion and clothing manufacturing.

REFERENCES

Ak, K. E., A. A. Kassim, J. H. Lim and J. Y. Tham (2018), "Learning Attribute Representations with Localization for Flexible Fashion Search," in *IEEE/CVF Conference on Computer Vision and Pattern Recognition*, Salt Lake City, UT, USA, pp. 7708–7717. https://doi.org/10.1109/CVPR.2018.00804.

Alsamarah, W., B. Younes and M. Yousef (2021), "Reducing Waste in Garment Factories by Intelligent Planning of Optimal Cutting Orders," *The Journal of The Textile Institute*, vol. 113, no. 9, pp. 1917–1925. https://doi.org/10.1080/00405000.2021.1956711.

Borji, A. (2022), "Generated Faces in the Wild: Quantitative Comparison of Stable Diffusion, Midjourney and Dall-e 2," *arXiv:2210.00586*. https://arxiv.org/abs/2210.00586.

Bouziri, A. and R. M'hallah (2007), "A Hybrid Genetic Algorithm for the Cut Order Planning Problem," in Okuno, H. G., Ali, M. (eds) *New Trends in Applied Artificial Intelligence*. IEA/AIE 2007. Lecture Notes in Computer Science, vol. 4570. Springer, Berlin, Heidelberg. https://doi.org/10.1007/978-3-540-73325-6_45

Cao, S., W. Chai, S. Hao, Y. Zhang, H. Chen and G. Wang (2024), "DiffFashion: Reference-Based Fashion Design With Structure-Aware Transfer by Diffusion Models," *IEEE Transactions on Multimedia*, vol. 26, pp. 3962–3975. https://doi.org/10.1109/TMM.2023.3318297.

Chakraborty, S., M. S. Hoque, N. Rahman Jeem, M. C. Biswas, D. Bardhan and E. Lobaton (2021), "Fashion Recommendation Systems, Models and Methods: A Review," *Informatics*, vol. 8, p. 49. https://doi.org/10.3390/informatics8030049

Chan, A. P., J. Fan and W. Yu (2003), "Men's Shirt Pattern Design Part II: Prediction of Pattern Parameters from 3D Body Measurements," *Sen'i Gakkaishi*, vol. 59, no. 8, pp. 328–333. Released on J-STAGE September 30, 2004. https://doi.org/10.2115/fiber.59.328, https://www.jstage.jst.go.jp/article/fiber/59/8/59_8_328/_article/-char/en.

Chang, A. A. et al. (2021), "Fashion Trend Forecasting Using Machine Learning Techniques: A Review," in Silhavy, R., Silhavy, P., Prokopova, Z. (eds) *Data Science and Intelligent Systems. CoMeSySo 2021. Lecture Notes in Networks and Systems*, vol. 231. Springer, Cham. https://doi.org/10.1007/978-3-030-90321-3_5.

Chen, H.-J., H.-H. Shuai and W.-H. Cheng (2023, May), "A Survey of Artificial Intelligence in Fashion," *IEEE Signal Processing Magazine*, vol. 40, no. 3, pp. 64–73. https://doi.org/10.1109/MSP.2022.3233449.

Chen, Y., Z. Wang, Y. Peng, Z. Zhang, G. Yu and J. Sun (2018 June 18–22), "Cascaded Pyramid Network for Multi-Person Pose Estimation," in *2018 IEEE Conference on Computer Vision and Pattern Recognition, CVPR 2018*, Salt Lake City, UT, USA, pp. 7103–7112. https://www.computer.org/csdl/proceedings-article/cvpr/2018/642000h103/17D45XuDNEl.

Cheng, Wen-Huang, Sijie Song, Chieh-Yun Chen, Shintami Chusnul Hidayati and Jiaying Liu (2021 June), "Fashion Meets Computer Vision: A Survey," *ACM Comput. Surv.*, vol. 54, no. 4, Article 72, p. 41. https://doi.org/10.1145/3447239.

Choi, W. et al. (2023), "Developing an AI-Based Automated Fashion Design System: Reflecting the Work Process of Fashion Designers," *Fashion and Textiles*, vol. 10, no. 39. https://doi.org/10.1186/s40691-023-00360-w.

Cui, Y. R., Q. Liu, C. Y. Gao and Z. Su (2018), "FashionGAN: Display Your Fashion Design Using Conditional Generative Adversarial Nets," *Computer Graphics Forum*, vol. 37, pp. 109–119. https://doi.org/10.1111/cgf.13552

Deldjoo, Y., Fatemeh Nazary, Arnau Ramisa, Julian McAuley, Giovanni Pellegrini, Alejandro Bellogin and Tommaso Di Noia (2023/2024 April), "A Review of Modern Fashion Recommender Systems," *ACM Computing Surveys*, vol. 56, no. 4, Article 87, p. 37. https://doi.org/10.1145/3624733.

Dong, H. et al. (2020), "Fashion Editing with Adversarial Parsing Learning," in *2020 IEEE/CVF Conference on Computer Vision and Pattern Recognition (CVPR)*, Seattle, WA, USA, pp. 8117–8125. https://doi.org/10.1109/CVPR42600.2020.00814.

Du, W., Y. Tang, S. Y. S. Leung, L. Tong, A. V. Vasilakos and F. Qian (2018 January), "Robust Order Scheduling in the Discrete Manufacturing Industry: A Multiobjective Optimization Approach," *IEEE Transactions on Industrial Informatics*, vol. 14, no. 1, pp. 253–264. https://doi.org/10.1109/TII.2017.2664080.

Fu, T., F. Li, Y. Zheng and R. Song (2021), "Process Learning of Robot Fabric Manipulation Based on Composite Reward Functions," in Liu, X. J., Nie, Z., Yu, J., Xie, F., Song, R. (eds) *Intelligent Robotics and Applications. ICIRA 2021. Lecture Notes in Computer Science*, vol. 13014. Springer, Cham. https://doi.org/10.1007/978-3-030-89098-8_16

Giri, C., S. Jain, X. Zeng and P. Bruniaux (2109), "A Detailed Review of Artificial Intelligence Applied in the Fashion and Apparel Industry," *IEEE Access*, vol. 7, pp. 95376–95396. https://doi.org/10.1109/ACCESS.2019.2928979.

Goenka et al. (2022), "FashionVLP: Vision Language Transformer for Fashion Retrieval with Feedback," in *IEEE/CVF Conference on Computer Vision and Pattern Recognition (CVPR)*, New Orleans, LA, USA, pp. 14085–14095. https://doi.org/10.1109/CVPR52688.2022.01371.

Gu, N., R. He and L. Yu (2024 February), "Learning to Unfold Garment Effectively Into Oriented Direction," *IEEE Robotics and Automation Letters*, vol. 9, no. 2, pp. 1051–1058. https://doi.org/10.1109/LRA.2023.3341763.

Gu, X., Fei Gao, Min Tan and Pai Peng (2020), "Fashion Analysis and Understanding With Artificial Intelligence," *Information Processing & Management*, vol. 57, no. 5, p. 102276. ISSN 0306-4573, https://doi.org/10.1016/j.ipm.2020.102276.

Günel, M., E. Erdem and A. Erdem (2018), "Language Guided Fashion Image Manipulation with Feature-Wise Transformations," *arXiv:1808.04000*. https://doi.org/10.48550/arXiv.1808.04000.

Guo, Z., Z. Zhu, Y. Li, S. Cao, H. Chen and G. Wang (2023), "AI Assisted Fashion Design: A Review," *IEEE Access*, vol. 11, pp. 88403–88415. https://doi.org/10.1109/ACCESS.2023.3306235.

Guo, Z. X. and W. K. Wong (2013), "Fundamentals of Artificial Intelligence Techniques for Apparel Management Applications," in *Optimizing Decision Making in the Apparel Supply Chain Using Artificial Intelligence (AI): From Production to Retail* (pp. 13–40). Elsevier Inc.

Guo, Z. X., W. K. Wong, S. Y. S. Leung, J. T. Fan and S. F. Chan (2013), "Optimizing Apparel Production Order Planning Scheduling Using Genetic Algorithms," in *Optimizing Decision Making in the Apparel Supply Chain Using Artificial Intelligence (AI): From Production to Retail* (pp. 55–80). Elsevier Inc.

Hu, Y., Liqiang Xiao, Yongkun Wang and Yaohui Jin (2018 October 29), "Fashion Pose Machine for Fashion Landmark Detection," in *Proc. SPIE 10836, 2018 International Conference on Image and Video Processing, and Artificial Intelligence*, 108360Y. https://doi.org/10.1117/12.2515278.

Huang, C. Q., J. K. Chen, Y. Pan, H. J. Lai, J. Yin and Q. H. Huang (2019, October), "Clothing Landmark Detection Using Deep Networks With Prior of Key Point Associations," *IEEE Trans.*

Cybern., vol. 49, no. 10, pp. 3744–3754. https://doi.org/10.1109/TCYB.2018.2850745, Epub 2018 Jul 12. PMID: 30004899.

Huang, J., R. Feris, Q. Chen and S. Yan (2015), "Cross-Domain Image Retrieval with a Dual Attribute-Aware Ranking Network," in *2015 IEEE International Conference on Computer Vision (ICCV)*, Santiago, Chile, pp. 1062–1070. https://doi.org/10.1109/ICCV.2015.12.

Huang, X., M. Y. Liu, S. Belongie and J. Kautz (2018), "Multimodal Unsupervised Image-to-Image Translation," in Ferrari, V., Hebert, M., Sminchisescu, C., Weiss, Y. (eds) *Computer Vision – ECCV 2018. Lecture Notes in Computer Science*, vol. 11207. Springer, Cham. https://doi.org/10.1007/978-3-030-01219-9_11.

Huang, Y., H. Shen, Y. Shi, W. Yang, W. Wang, R. Wan and L. Dodd (2024), "Artificial Intelligence-Based Precise Prediction of Anthropometric Data for Female Garment Pattern-Making," *The Journal of The Textile Institute*, pp. 1–9. https://doi.org/10.1080/00405000.2024.2352183.

Jack Tech (2023), *Jack A5E-A Brochure*. https://www.jacksew.com/content/docs/jack/JK-A5E-A/Jack%20A5E-A%20Brochure.pdf.

Jiang, Shuhui and Yun Fu (2017), "Fashion Style Generator," in *Proceeding of 26th International Joint Conf. Artificial Intelligence (IJCAI)*, pp. 3721–3727.

Jiang, Shuhui, J. Li and Y. Fu (2021), "Deep Learning for Fashion Style Generation," *IEEE Transactions on Neural Networks and Learning Systems*, vol. 33, no. 9, pp. 4538–4550.

Kiapour, M. H., X. Han, S. Lazebnik, A. C. Berg and T. L. Berg (2015), "Where to Buy It: Matching Street Clothing Photos in Online Shops," in *2015 IEEE International Conference on Computer Vision (ICCV)*, Santiago, Chile, pp. 3343–3351. https://doi.org/10.1109/ICCV.2015.382. https://ieeexplore.ieee.org/abstract/document/7410739

Kouslis, E. et al. (2024), "AI in Fashion: A Literature Review," *Electronic Commerce Research*. https://doi.org/10.1007/s10660-024-09872-z.

Koustoumpardis, P. N. and N. A. Aspragathos (2011), "Robotized Sewing of Fabrics Based on a Force Neural Network Controller," in Jeschke, S., Liu, H., Schilberg, D. (eds) *Intelligent Robotics and Applications. ICIRA 2011. Lecture Notes in Computer Science*, vol. 7101. Springer, Berlin, Heidelberg. https://doi.org/10.1007/978-3-642-25486-4_49.

Ku, S., H. Choi, H.-Y. Kim and Y.-L. Park (2023, September), "Automated Sewing System Enabled by Machine Vision for Smart Garment Manufacturing," *IEEE Robotics and Automation Letters*, vol. 8, no. 9, pp. 5680–5687. https://doi.org/10.1109/LRA.2023.3300284.

Landim, A. R. D. B., A. M. Pereira, T. Vieira, E. de B. Costa, J. A. B. Moura, V. Wanick and E. Bazaki (2021), "Chatbot Design Approaches for Fashion E-Commerce: An Interdisciplinary Review," *International Journal of Fashion Design, Technology and Education*, vol. 15, no. 2, pp. 200–210. https://doi.org/10.1080/17543266.2021.1990417.

Liang, X., L. Lin, W. Yang, P. Luo, J. Huang and S. Yan (2016 June), "Clothes Co-Parsing Via Joint Image Segmentation and Labelling with Application to Clothing Retrieval," *IEEE Transactions on Multimedia*, vol. 18, no. 6, pp. 1175–1186. https://doi.org/10.1109/TMM.2016.254298.

Liang, X. et al. (2015), "Human Parsing with Contextualized Convolutional Neural Network," in *2015 IEEE International Conference on Computer Vision (ICCV)*, Santiago, Chile, pp. 1386–1394. https://doi.org/10.1109/ICCV.2015.163.

Liu, J. and Hong Lu (2019), "Deep Fashion Analysis with Feature Map Upsampling and Landmark-Driven Attention," in *Computer Vision – ECCV 2018 Workshops: Munich, Germany, September 8–14, 2018, Proceedings, Part III* (pp. 30–36). Springer-Verlag, Berlin, Heidelberg. https://doi.org/10.1007/978-3-030-11015-4_4

Liu, K., J. Wang, E. Kamalha, V. Li and X. Zeng (2017), "Construction of a Prediction Model for Body Dimensions Used in Garment Pattern Making Based on Anthropometric Data

Learning," *The Journal of The Textile Institute*, vol. 108, no. 12, pp. 2107–2114. https://doi.org/10.1080/00405000.2017.1315794

Liu, L., Haijun Zhang, Yuzhu Ji and Q. M. Jonathan Wu (2019), "Toward AI Fashion Design: An Attribute-GAN Model for Clothing Match," *Neurocomputing*, vol. 341, pp. 156–167. ISSN 0925-2312, https://doi.org/10.1016/j.neucom.2019.03.011.

Liu, S., X. Liang, L. Liu, X. Shen, J. Yang, C. Xu, L. Lin, X. Cao and S. Yan (2015 June), "Matching-CNN Meets KNN: Quasi-Parametric Human Parsing," in *Proc. IEEE Conf. Comput. Vis. Pattern Recognit. (CVPR)*, pp. 1419–1427. https://arxiv.org/abs/1504.01220.

Liu, Z., P. Luo, S. Qiu, X. Wang and X. Tang (2016a June), "Deepfashion: Powering Robust Clothes Recognition and Retrieval with Rich Annotations," in *2016 IEEE Conference on Computer Vision and Pattern Recognition (CVPR)*, pp. 1096–1104.

Liu, Z., S. Yan, P. Luo, X. Wang and X. Tang (2016b October), "Fashion Landmark Detection in the Wild," in *European Conference on Computer Vision (ECCV)*. https://arxiv.org/abs/1608.03049.

Lomas, Q., A. Goti Elordi, A. Almeida Escondrillas and D. Lopez De Ipiña Gonzalez De Artaza (2021), "A Systematic Literature Review of Artificial Intelligence in Fashion Retail B2C," in *2021 6th International Conference on Smart and Sustainable Technologies (SpliTech), Bol and Split*, Croatia, pp. 1–6. https://doi.org/10.23919/SpliTech52315.2021.9566467.

Luce, L. 2019. "Artificial Intelligence for Fashion, How AI is Revolutionizing the Fashion Industry," Apress Berkeley, CA. https://doi.org/10.1007/978-1-4842-3931-5

Ma, S. and X. Yan (2022 May 10), "Application of BP Neural Networks in Garment Pattern Design System," *Comput. Intell. Neurosci.*, p. 8766137. https://doi.org/10.1155/2022/8766137. PMID: 35592711, PMCID: PMC9113887.

Matas, J., S. James and A. J. Davison (2018), "Sim-to-Real Reinforcement Learning for Deformable Object Manipulation," in *2nd Conference on Robot Learning, Proceedings of Machine Learning Research*, vol. 87, pp. 734–743. https://proceedings.mlr.press/v87/matas18a.html.

Mok, P. Y. et al. (2013), "Intelligent Production Planning for Complex Garment Manufacturing," *J. Intell. Manuf.*, vol. 24, pp. 133–145. https://doi.org/10.1007/s10845-011-0548-y

Mok, P. Y. et al. (2007), "Optimisation of Fault-Tolerant Fabric-Cutting Schedules Using Genetic Algorithms and Fuzzy Set Theory," *European Journal of Operational Research*, vol. 177, no. 3, pp. 1876–1893, ISSN 0377-2217. https://doi.org/10.1016/j.ejor.2005.12.021.

Mypati, O. et al. (2023), "A Critical Review on Applications of Artificial Intelligence in Manufacturing," *Artif. Intell. Rev.*, vol. 56, no. Suppl 1, pp. 661–768. https://doi.org/10.1007/s10462-023-10535-y.

Ng, R. (2004), "Garment Pattern Design with Artificial Neural Network and Fuzzy Logic," in *Paper Presented at International Fashion Culture Festival*, Shanghai (cited in Yang et al., 2007).

Noor, A., M. A. Saeed, T. Ullah, Z. Uddin and R. M. W. Ullah Khan (2021), "A Review of Artificial Intelligence Applications in Apparel Industry," *The Journal of The Textile Institute*, vol. 113, no. 3, pp. 505–514. https://doi.org/10.1080/00405000.2021.1880088.

Oh, J. and S. Kim (2023), "Automatic Generation of Parametric Patterns from Grading Patterns Using Artificial Intelligence," *International Journal of Clothing Science and Technology*, vol. 35, no. 6, pp. 889–903. https://doi.org/10.1108/IJCST-07-2022-0102.

Patrick, C. H., S. Ng Frency and C. Chan Keith (2000), "A Study of the Roll Planning of Fabric Spreading Using Genetic Algorithms," *International Journal of Clothing Science and Technology*, vol. 12, no. 1, pp. 50–62. https://doi.org/10.1108/09556220010313832.

Pillarisetty, R. and P. Mishra (2022), "A Review of AI (Artificial Intelligence) Tools and Customer Experience in Online Fashion Retail," *International Journal of E-Business Research (IJEBR)*, vol. 18, no. 2, pp. 1–12. http://doi.org/10.4018/IJEBR.294111.

Ribeiro, R. et al. (2021), "An Intelligent Decision Support System for Production Planning in Garments Industry," in Yin, H. et al. (eds) *Intelligent Data Engineering and Automated Learning – IDEAL 2021. IDEAL 2021. Lecture Notes in Computer Science*, vol. 13113. Springer, Cham. https://doi.org/10.1007/978-3-030-91608-4_37.

Ruan, T., T. Liu, Z. Huang, Y. Wei, S. Wei and Y. Zhao (2019), "Devil in the Details: Towards Accurate Single and Multiple Human Parsing," *Proceedings of the AAAI Conference on Artificial Intelligence*, vol. 33, no. 1, pp. 4814–4821. https://doi.org/10.1609/aaai.v33i01.33014814.

Särmäkari, N. and A. Vänskä (2021), "'Just Hit a Button!' – Fashion 4.0 Designers as Cyborgs, Experimenting and Designing with Generative Algorithms," *International Journal of Fashion Design, Technology and Education*. https://doi.org/10.1080/17543266.2021.1991005

Shenshen, Z. (2019), *AI Designer Takes Applause on Shanghai Catwalk*. https://www.shine.cn/biz/tech/1904243580/ (last visited 29 September 2024).

Stylios, G. K. (2013). Intelligent sewing systems for garment automation and robotics. Chapter 7. In I. Jones & G. K. Stylios (Eds.), *Joining textiles: Principles and applications* (pp. 208–221). Woodhead Publishing Ltd. (Woodhead Publishing Series in Textiles No. 110). https://doi.org/10.1533/9780857093967.1.208.

Tang, Y., S. Li, Y. Ye and J. Ma (2019), "Spatial-Aware Non-Local Attention for Fashion Landmark Detection," in *2019 IEEE International Conference on Multimedia and Expo (ICME)*, Shanghai, China, pp. 820–825. https://doi.org/10.1109/ICME.2019.00146, https://ieeexplore.ieee.org/abstract/document/8784817.

Tsao, Y. C., C. H. Hung and T. L. Vu (2021), "Hybrid Heuristics for Marker Planning in the Apparel Industry," *Arab. J. Sci. Eng.*, vol. 46, pp. 10077–10096. https://doi.org/10.1007/s13369-020-05210-1.

Tsao, Y.-C., C.-C. Liu, P.-R. Chen and T.-L. Vu (2024), "Design Heuristics for Cut Order Planning Problems in the Apparel Industry," *International Journal of Clothing Science and Technology*, vol. 36, no. 4, pp. 662–686. https://doi.org/10.1108/IJCST-07-2022-0087.

Tsao, Y. C., Thuy-Linh Vu and Lu-Wen Liao (2020), "Hybrid Heuristics for the Cut Ordering Planning Problem in Apparel Industry," *Computers & Industrial Engineering*, vol. 144, p. 106478. ISSN 0360-8352, https://doi.org/10.1016/j.cie.2020.106478.

Wang, W., Y. Xu, J. Shen and S.-C. Zhu (2018), "Attentive Fashion Grammar Network for Fashion Landmark Detection and Clothing Category Classification," in *2018 IEEE/CVF Conference on Computer Vision and Pattern Recognition*, Salt Lake City, UT, USA, pp. 4271–4280. https://doi.org/10.1109/CVPR.2018.00449.

Wang, Z., Jianping Wang, Xianyi Zeng, Xuyuan Tao, Yingmei Xing and Pascal Bruniaux (2021), "Construction of Garment Pattern Design Knowledge Base Using Sensory Analysis, Ontology and Support Vector Regression Modeling," *International Journal of Computational Intelligence Systems*, vol. 14, no. 1. https://doi.org/10.2991/ijcis.d.210608.002.

Wong, W. K. (2013), "Understanding Key Decision Points in the Apparel Supply Chain," in *Optimizing Decision Making in the Apparel Supply Chain Using Artificial Intelligence (AI): From Production to Retail* (pp. 1–12). Elsevier Inc.

Wong, W. K. and C. K. Chan (2001), "An Artificial Intelligence Method for Planning the Clothing Manufacturing Process," *Journal of the Textile Institute*, vol. 92, no. 2, pp. 168–178. https://doi.org/10.1080/00405000108659606.

Wong, W. et al. (2005), "Optimization of Manual Fabric-Cutting Process in Apparel Manufacture Using Genetic Algorithms," *International Journal of Advanced Manufacturing Technology*, vol. 27, pp. 152–158. https://doi.org/10.1007/s00170-004-2161-0

Wong, W. K. et al. (2006), "Genetic Optimization of JIT Operation Schedules for Fabric-Cutting Process in Apparel Manufacture," *Journal of Intelligent Manufacturing*, vo. 113, pp. 341–354. https://doi.org/10.1007/s10845-005-0007-8

Wong, W. K., C. K. Kwong, P. Y. Mok and W. H. Ip (2013d), "Optimizing Fabric Spreading and Cutting Schedules in Apparel Production Using Genetic Algorithms and Fuzzy Set Theory," in *Optimizing Decision Making in the Apparel Supply Chain Using Artificial Intelligence (AI): From Production to Retail* (pp. 132–152). Elsevier Inc.

Wong, W. K. and S. Y. S. Leung (2008), "Genetic Optimization of Fabric Utilization in Apparel Manufacturing," *International Journal of Production Economics*, vol. 114, no. 1, pp. 376–387. ISSN 0925-5273, https://doi.org/10.1016/j.ijpe.2008.02.012.

Wong, W. K. and S. Y. S. Leung (2013), "Optimizing Cut Order Planning in Apparel Production Using Evolutionary Strategies," in *Optimizing Decision Making in the Apparel Supply Chain Using Artificial Intelligence (AI): From Production to Retail* (pp. 81–105). Elsevier Inc.

Wong, W. K., P. Y. Mok and S. Y. S. Leung (2013b), "Optimizing Apparel Production Systems Using Genetic Algorithms," in *Optimizing Decision Making in the Apparel Supply Chain Using Artificial Intelligence (AI): From Production to Retail* (pp. 153–169). Elsevier Inc.

Wong, W. K., X. X. Wang and Z. X. Guo (2013c), "Optimizing Marker Planning in Apparel Production Using Evolutionary Strategies and Neural Networks," in *Optimizing Decision Making in the Apparel Supply Chain Using Artificial Intelligence (AI): From Production to Retail* (pp. 106–131). Elsevier Inc.

Wong, W. K., X. H. Zeng and K. F. Au (2013a), "Selecting the Location of Apparel Manufacturing Plants Using Neural Networks," in *Optimizing Decision Making in the Apparel Supply Chain Using Artificial Intelligence (AI): From Production to Retail* (pp. 41–54), Elsevier Inc.

WSJ (2015), https://www.youtube.com/watch?v=BA96-WX-oXc.

Wu, X. and L. Li (2024), "An Application of Generative AI for Knitted Textile Design in Fashion," *The Design Journal*, vol. 27, no. 2, pp. 270–290. https://doi.org/10.1080/14606925.2024.2303236

Xian, W., P. Sangkloy, V. Agrawal, A. Raj, J. Lu, C. Fang, F. Yu and J. Hays (2018 June), "TextureGAN: Controlling Deep Image Synthesis with Texture Patches," in *Proceeding of 2018 IEEE/CVF Conference on Computer Vision and Pattern Recognition (CVPR)*, pp. 8456–8465. https://doi.org/10.1109/CVPR.2018.00882.

Yan, H., Haijun Zhang, Linlin Liu, Dongliang Zhou, Xiaofei Xu, Zhao Zhang and Shuicheng Yan (2023a), "Toward Intelligent Design: An AI-Based Fashion Designer Using Generative Adversarial Networks Aided by Sketch and Rendering Generators," *Trans. Multi.*, vol. 25, pp. 2323–2338. https://doi.org/10.1109/TMM.2022.3146010.

Yan, H., Haijun Zhang, Jianyang Shi and Jianghong Ma (2023b May), "Texture Brush for Fashion Inspiration Transfer: A Generative Adversarial Network With Heatmap-Guided Semantic Disentanglement," *IEEE Trans. Cir. and Sys. for Video Technol.*, vol. 33, no. 5, pp. 2381–2395. https://doi.org/10.1109/TCSVT.2022.3224190.

Yan, H., Haijun Zhang, Jianyang Shi, Jianghong Ma and Xiaofei Xu (2023c May). "Toward Intelligent Fashion Design: A Texture and Shape Disentangled Generative Adversarial Network," *ACM Trans. Multimedia Comput. Commun. Appl.*, vol. 19, no. 3, Article 107, p. 23. https://doi.org/10.1145/3567596.

Yan, S., Ziwei Liu, Ping Luo, Shi Qiu, Xiaogang Wang and Xiaoou Tang (2017), "Unconstrained Fashion Landmark Detection via Hierarchical Recurrent Transformer Networks," in *Proceedings of the 25th ACM international Conference on Multimedia (MM '17)* (pp. 172–180). Association for Computing Machinery, New York, NY, USA. https://doi.org/10.1145/3123266.3123276.

Yang, Y., W. Zhang and C. Shan (2007), "Investigating the Development of Digital Patterns for Customized Apparel," *International Journal of Clothing Science and Technology*, vol. 19, no. 3–4, pp. 167–177. https://doi.org/10.1108/09556220710741632.

Yu, F. and V. Koltun (2015), "Multi-Scale Context Aggregation by Dilated Convolutions," in *Published as a Conference Paper at ICLR 2016, Computer Vision and Pattern Recognition (cs.CV), arXiv:1511.07122.* https://arxiv.org/abs/1511.07122.

Yu, W., X. Liang, K. Gong, C. Jiang, N. Xiao and L. Lin (2019), "Layout-Graph Reasoning for Fashion Landmark Detection," in *2019 IEEE/CVF Conference on Computer Vision and Pattern Recognition (CVPR)*, Long Beach, CA, USA, pp. 2932–2940. https://doi.org/10.1109/CVPR.2019.00305, https://ieeexplore.ieee.org/document/8953702.

Zhang, F., Mingliang Xu and Changsheng Xu (2022 May), "Tell, Imagine, and Search: End-to-end Learning for Composing Text and Image to Image Retrieval," *ACM Trans. Multimedia Comput. Commun. Appl.*, vol. 18, no. 2, Article 59. https://doi.org/10.1145/3478642.

Zhang, Yanbo and Chuanlan Liu (2024), "Unlocking the Potential of Artificial Intelligence in Fashion Design and E-Commerce Applications: The Case of Midjourney," *Journal of Theoretical and Applied Electronic Commerce Research*, vol. 19, no. 1, pp. 654–670. https://doi.org/10.3390/jtaer19010035.

Zhang, Yungang, C. Zhang and F. Du (2019), "A Brief Review of Recent Progress in Fashion Landmark Detection," in *2019 12th International Congress on Image and Signal Processing, BioMedical Engineering and Informatics (CISP-BMEI)*, Suzhou, China, pp. 1–6. https://ieeexplore.ieee.org/document/8966051.

COMPANIES

https://3dexperiencelab.3ds.com.
https://www.sewts.com.

2 Co-Ideation with Generative Artificial Intelligence (GAI) to Develop Fashion Textiles

Si Eun Kim and Se Jin Kim

2.1 INTRODUCTION

In 2023, the world's first AI Fashion Week was held, featuring generative artificial intelligence (AI). Participants from various backgrounds conducted online exhibitions showcasing images created by AI. Professional designers and the general public took part in Fashion Week. Generative AI is utilised in numerous fields, including fashion, and is expected to see widespread use due to its low entry barrier and versatility. Research on fashion design using this technology is still in its early stages. Text-to-image generation AI can create images based on a simple prompt input, allowing for interactions between the AI and the user to generate new images. However, there are issues with copyright, information transmission errors, and cultural biases in the processes and results (Kim 2024). As interest in generative AI within the fashion industry continues to grow, recent research has actively explored various aspects, including proposing applicable AI models (Liu et al. 2019; Jeon et al. 2021), analysing processes from the perspective of the fashion design system (Lee 2022; Choi et al. 2023), and presenting design development case studies (Jung and Suh 2024). However, research on design development and ideation through interactions between generative AI and designers remains in its infancy, with only a limited number of studies addressing cultural heritage while considering sociocultural issues.

This study aims to explore the interaction between generative AI and designers in the collaborative ideation process for the development of fashion textiles inspired by traditional Korean culture and heritage. The study includes three main components: 1) literature research on Gaya culture and artifacts and the generation of collaborative ideas using AI, 2) design development through collaborative ideas with AI, and 3) a discussion of the process and implications of using AI for collaborative ideation. The results of this study will inform design development and training using generative AI, which is becoming more widely used in the industry.

This study focuses on the ancient Gaya tombs, a traditional Korean cultural site that has been listed as a UNESCO World Heritage Site for 23 years. The Gaya tombs

 DOI: 10.1201/9781003594482-3

offer significant evidence of East Asia's ancient civilisation and possess outstanding universal value (OUV) (Hee 2023). It is also a culture that is not well-known despite its importance. Understanding the role and potential of an AI platform in developing collaborative design using generative AI is considered appropriate. Furthermore, the use of generative AI technology in design development can minimise energy and resource consumption prior to physical production while facilitating the generation of diverse ideas. Highlighting this process can contribute to design education by showcasing the potential of new digital tools and offering valuable insights for industrial applications.

2.2 BACKGROUND

2.2.1 Collaborative Ideation (Co-Ideation) with AI

Ideation involves generating, refining, and sharing ideas (Jonson 2005). This cognitive process guides the knowledge stored in the designer's memory, obtained from direct and indirect experiences, toward a solution for a specific design problem (Finneran and Zhang 2005). The design stage involves various cognitive activities. During this phase, designers use a variety of information, including photos and text. In the fashion field, where visual thinking is crucial, photos serve as a significant source of inspiration and play a vital role in idea development. Image-generative AI enables users to obtain desired image outcomes by interacting with AI agents using text or images.

The advent of AI heralds a change in the fashion design process. Research on the application of AI, especially GAI, to fashion design development is actively underway.

Guo et al. (2023) reviewed the integration of enhanced personalisation and seamless multimodal interfaces in the field of fashion design and recommendation and confirmed that it can provide highly personalised fashion product recommendations and effectively communicate design ideas between designers and users. Choi et al. (2023) collected and analysed existing cases of AI-based clothing design tools to identify similarities and differences between the clothing development process of human designers and existing AI-based clothing design tools, and proposed an AI-based clothing development system. This innovative approach neglects the critical sociocultural context and the designer's creative ideation in the fashion design process. To enhance satisfaction with design outcomes, it is essential to consider the quality of sources influencing design development and the creativity of the designer (Rostamzadeh et al. 2018). This study emphasises incorporating interaction with GAI during the designer's ideation process.

This process may also lead to the generation of new ideas (Kim 2024). In this study, we define the process of problem exploration, goal setting, information collection, related image creation, and design development as ideation for fashion textile development. The use of generative AI tools during the information collection and related image creation stages is considered collaborative ideation.

2.2.2 Gaya Culture

Gaya was a nation that emerged in the southern region of present-day Korea from year 42 to 532. During its time, Gaya vied for dominance on the Korean Peninsula alongside three kingdoms – Goguryeo, Baekje, and Silla – before fading into obscurity. Gaya remains a mysterious ancient kingdom in Korea, with more hidden information than revealed.

Geumgwan Gaya, once known as the ancient name of the Gimhae region in southern Korea, played a significant role in the late Silla and early Goryeo period. Its name translates to 'Geumgwan country among the Gaya confederation'. Though evidence of Gaya's existence is limited due to its short lifespan, pottery and metal relics unearthed from the Gaya cultural area showcase intricate decorations coexisting with modern organic and geometric forms. Essentially, the formative characteristics of Gaya relics reflect an exquisite harmony of soft curves and relief, embodying sophistication and elegance (Byun 2018).

Among the relics of Geumgwan Gaya, pottery expresses the largest number of patterns. However, the number of excavated relics is limited. As a result, although there is great potential to use them as fashion design materials, this limitation hinders design development. Meanwhile, a few artifacts that are more clearly preserved include the gold crowns. Gaya's gold crown measures 11.5 cm in height, 20.7 cm in diameter, and 3.6 cm in width. The elaborate and refined shapes of the ornaments give a sense of unity, the soft curves create a sense of rhythm, the symmetrical patterns create a sense of balance, and the pointed jade decorations create harmony. The crown has a wide rim, an onion-shaped blade of grass, and four upright decorations. This gold crown from the peak of Gaya's civilisation is considered a masterpiece that showcases exceptional artistic vision and craftsmanship (Kim 2017).

On the other hand, patterns are a reflection of consciousness, a product of mental activity, and a result of creative beautification activities. In groups that enjoy patterns, they have the characteristics of promised symbols. Korean traditional patterns are representative elements that express tradition and often contain symbolic meanings (Chang 1998). Although it is still difficult to find Gaya relics with patterns applied, this study aims to recreate them as modern pattern designs using the forms of relics that can be confirmed. This can symbolise and express the cultural identity of the Gaya period and relics, and it will provide the possibility of developing new forms and ideas.

Design development using Gaya culture, which is not well-known and lacks content, can be challenging in the design process. We anticipate results that enhance ideas through collaborative ideation with the help of generative AI.

2.2.3 Cultural Heritage in Textile and Fashion Design

Previous studies on fashion design and material development using tangible cultural heritage have shown that traditional cultural heritage can be an excellent source of ideas and educational resources for fashion design that appeals to new generations. Cultural heritage plays an important role in the design and evelopment of fashion cultural products (Kim and Oh 2012). It has significance in that it developed designs targeting the younger generation using Jikji, the world's oldest metal type. The results confirmed that traditional culture as a material can provide newness that can appeal to the younger generation of Korea who want modern and interesting designs, while also contributing to promoting and increasing the value of Korean cultural resources. Csanák (2019) explored the possibility of reinterpreting national traditions, crafts, and traditional folk motifs in fashion and textile design through student project work. The researcher explained that such projects carried out through student exchange programmes with international students, can contribute to the issue of sustainability by providing students with the opportunity to gain essential collection development and experience while also providing students from other countries with the opportunity to learn about foreign cultures, traditions, and heritage (Csanák 2019).

In addition, as a study on applying AI to design development, Xu et al. (2021) used a neural network-enhanced interactive genetic algorithm (IGA) to integrate intangible cultural heritage and traditional craftsmanship into modern fashion design. The results of this study suggest the positive effects of applying intelligent algorithms to fashion design and cultural heritage promotion. However, AI tools cannot completely replace the creativity and sensibility of human designers, and it may be difficult to accurately reflect complex cultural contexts. Therefore, research on using AI combined with human sensibility and creativity is needed.

2.3 METHODS

This study utilised literature on Gaya culture and textile design development with the help of generative AI. MidJourney served as the platform for generative AI. It is a GAN-based model that creates images based on input text or images, and functions through the discord server, an online community. Another similar platform is DALL.E 2, which is more stylish and offers detailed commands, making it popular among illustrators and for photo production. In contrast, MidJourney is preferred for its suitability in the design field and its ability to provide a variety of prompts and results for users to check.

In this study, we chose wedding dresses as garments for textile design. The wedding dress, inspired by Queen Victoria's attire in 1840, is worn by the bride on her wedding day and includes a white dress, veil, corsage, and accessories (Hong and Lee 2012). Wedding dresses typically highlight a woman's beautiful silhouette and capture attention through various elements or designs (Lee and Kwon 2023). They tend to evolve at a slower pace compared to other fashion trends, mainly due to restrictions in colour choices and traditional styles. However, there is a growing demand for change influenced by current times, trends, and the diversification of preferences.

Since wedding dresses serve a universal purpose and are widely used items, communication with AI was considered easy for collaboration. First, the distinguishing features of Gaya relics were identified through research on Gaya culture. Based on photographs of gold crown artifacts, motifs were extracted and simplified using Adobe Illustrator, considering the formative characteristics of Gaya relics. These motifs were slightly modified to create the initial design elements (Figure 2.1). Second, the requirements for developing wedding dress designs were determined. In the image-creation stage, a pattern motif inspired by Gaya relics was created, and design

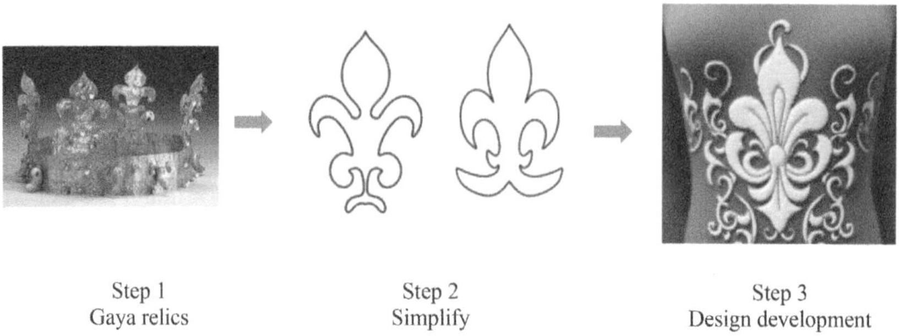

FIGURE 2.1 The development process of textile pattern design [Step 1: Gaya Relics (source: Leeum Museum (n.d.); Step 2: Simplify (image created by the authors using Adobe Illustrator); Step 3: Design development (image created by the authors using Midjourney)].

outcomes were generated using the MidJourney platform. During this stage, relevant images, motifs, and specific prompts were incorporated into the platform. The interaction was iterative, involving adjustments to prompts or inputting additional images until the desired design results were achieved. Finally, textile design was completed using illustration software and the MidJourney platform, effectively integrating the gathered information and design outputs.

2.4 RESULTS

Proposals for pattern design development and the creation of four distinct wedding dresses are outlined. The production process began with an analysis of visual resources related to Gaya culture, focusing on the embellishments of the gold crown and the motif of the onion-shaped blade of grass. These motifs were simplified using Illustrator to create basic design data. Using MidJourney's features to merge multiple images, prompts such as 'gold border' and 'pattern' were added to generate patterned imagery. Additional prompts related to 'wedding dress' were then included, resulting in iterative refinement until satisfactory patterns and wedding dress designs were achieved. This process produced a total of 18 results, from which four were selected based on three criteria: 1) the organic, geometric, splendid, harmonious, incongruous, rhythmic, dynamic, and balanced characteristics of Gaya relics; 2) suitability for wedding dresses; and 3) feasibility for production.

The design results are as follows: The first design featured a modified onion-shaped grass leaf motif attached to the top of the gold crown, capturing the splendour and classicism of the original shape (Figure 2.2). This dress incorporates simplified patterns inspired by the forms and structures observed in artefacts. The design embodies a sophisticated harmony of classic opulence and modern elegance, reflecting a sense of luxury. The pattern features curved designs inspired by the onion-shaped motifs observed in Gaya artefacts. These designs are embroidered in gold onto an A-line wedding dress, capturing Gaya artefacts' ornate and graceful qualities. The intricate and finely detailed craftsmanship of the motifs lends an air of

FIGURE 2.2 Design 1. (image created by the authors using MidJourney)

formal sophistication and luxury to the wedding attire. The second design presented a more delicate and sophisticated interpretation by including a woman's body silhouette, elegantly highlighting the symmetrical patterns on both sides to convey a sense of balance (Figure 2.3). This design emphasises a graceful silhouette with a fitted waist, featuring a cut-out detail at the back of the dress. Patterns extending below the waist enhance the three-dimensional effect, maximising the beauty of the feminine silhouette. The developed pattern achieves a harmonious balance through the uniformity of repetitive motifs, the symmetrical order, and the rhythmic appeal of geometric forms.

The third design showcased the dynamism and vitality inspired by the solemnity and grandeur of the gold crown, embodying the gentle yet strong identity of Gaya culture (Figure 2.4). This design was derived by combining images of Gaya tombs and golden crowns using MidJourney, resulting in a unique dress concept. The harmony between the floral background and the embroidery on the dress conveys purity and sophistication. Applying crown motifs as impactful accents on the voluminous, flowing dress adds a modern touch. The symmetrical pattern design provides balance and stability, while its bold and expansive arrangement reflects the majesty and grandeur of traditional royal heritage. The final design reflected the delicacy and symmetry of the gold crown, incorporating motifs that repeated on both sides to create a rhythmic silhouette of the woman's body and express harmonious proportions (Figure 2.5). This design is intended to highlight traditional elegance and refined

FIGURE 2.3 Design 2. (image created by the authors using MidJourney)

FIGURE 2.4 Design 3. (image created by the authors using MidJourney)

FIGURE 2.5 Design 4. (image created by the authors using MidJourney)

sophistication. It features an A-line silhouette that beautifully accentuates the bride's figure and incorporates patterns positioned to draw attention towards the body's centre. The contrast between the silhouette and the placement of the patterns is rendered harmonious and graceful, adding a sense of classic and distinguished luxury. These designs incorporate classic shapes that emphasise the female human body silhouette preferred by Korean women, Gaya gold crown motifs, gold embroidery, and a sense of harmony that blends elegance and rigidity in a unique way, achieving a sense of rhythm and balance.

These findings affirm the potential of Gaya relics to serve as unique and distinctive aesthetic elements in developing patterns for modern fashion and wedding dress design. While this study presented four final designs, the design development process using the Generative AI (GAI) platform yielded countless outcomes. This demonstrates the platform's efficiency and effectiveness in fostering creativity, particularly given the time and financial resources traditionally required for idea generation and sample production. Consequently, interaction with GAI provides valuable inspiration during the often-ambiguous initial stages of design ideation.

2.5 DISCUSSION AND CONCLUSION

Amid globalisation, the importance of local culture is increasing, and preserving and maintaining regional identity is a vital challenge at this time. Efforts have been made to promote a country's cultural values on a global scale. Contemporary fashion trends strive to reinterpret traditional materials with a modern sensibility, incorporating them into fashion in unique and distinctive ways. Traditional motifs, as historical assets passed down through generations, serve as aesthetic inspirations. Developing designs based on these motifs makes it possible to create fashion items with significant artistic value. Furthermore, there has been a growing interest in generative AI in the fashion industry. As a result, we have explored the possibility of using text-to-image generation AI based on Gaya relics, which are not well-known and lack sufficient information, but are recognised for their cultural value. Through collaborative ideation with generative AI, we have developed a wedding dress design pattern.

This process has several implications. First, collaborative work with AI in design development has proven to be advantageous, allowing for the generation of a significant number of results quickly despite limited initial information. Second, a variety of simulations close to reality are possible, leading to reduced time, materials, and energy required for sample production, as well as a decrease in errors during trial-and-error work. In fact, it only took just over an hour to obtain 18 primary results using AI, showcasing its efficiency in modern fashion systems that demand quick results in a short timeframe. However, as discussed later, there are also concerns regarding its effectiveness.

There are negative aspects to consider. First, the AI platform we used is popular in the art field but lacks knowledge of recent trends, making it unsuitable for the fashion industry, which requires future-oriented responses. While our study focused on wedding dresses, which are not heavily impacted by trends, limitations may arise in developing women's clothing that demands up-to-date fashion information. This highlights the need for specialised AI models in the fashion sector. Recent research on developing AI models for fashion design has been actively pursued (Liu et al. 2019; Yan et al. 2023). However, there is a growing need to develop models that incorporate human emotions and sociocultural contexts to enhance their relevance and application.

Second, due to a lack of expertise in fashion, subtle design differences in clothing were overlooked, leading to similar final results. This emphasises the importance of caution and specific guidelines when providing input. Our task relied on specific information and prior knowledge, but individuals without relevant skills may encounter misinformation or cultural biases in AI. In particular, related prior research by Vartiainen and Tedre (2023) explains that creating craftwork using generative AI can lead to issues like copyright infringement, cultural appropriation, and data privacy concerns. Kizilcec and Lee (2009) caution that AI algorithms trained on biased data can perpetuate and exacerbate these biases, resulting in discriminatory outcomes such as biased grading or course recommendations. Consequently, it is crucial to continuously monitor AI models, train them responsibly, and adopt measures to mitigate bias, thereby ensuring equity in educational contexts (Elmourabit et al. 2024). Our research on cultural heritage has successfully addressed these issues, but it is important not to ignore similar challenges. Therefore, parallel educational efforts for digital literacy are needed in the field of DL.

This study used collaborative ideation with AI to create clothing designs inspired by cultural heritage and showcased the process. The findings indicate that collaborative ideation with AI promotes divergent thinking essential for designers. Generative AI's main function is reducing trial and error by generating a variety of ideas and tasks. The results of this study suggest improved work efficiency in a short time frame. However, it remains unclear whether these results contribute to enhancing divergent thinking in design. There are several limitations to this study, such as the focus solely on developing motifs and graphic patterns in fashion textile design, overlooking texture and material mixing ratios. Additionally, the feedback from the intended users (i.e. brides-to-be) was not taken into account. While this was an exploratory study restricting the use of generative AI to design development, future research could validate the creativity, feasibility, and efficiency of the concept by implementing it with students or designers and comparing results before and after. We recommend conducting experiments and qualitative and quantitative research to delve deeper into the topic. Furthermore, exploring user perceptions and preferences of design outcomes utilising AI would be beneficial. On the other hand, although it did not lead to actual production, the development of pattern designs using new digital tools holds significant value in advancing design development methods in the fashion industry.

2.6 ACKNOWLEDGEMENTS

The authors declare no potential conflict of interest. The original contributions presented in the study are included in the article.

REFERENCES

Byun, Sung-Tae. 2018. "A Study on the Development of the Pattern and the Necktie Design Utilizing the Gaya Relics." *The Journal of the Convergence on Culture Technology* 4 (4): 21–30. https://doi.org/10.17703/JCCT.2018.4.4.21.

Chang, Soo Kyung. 1998. "A Study on the Figuration of Korean Traditional Pattern Images." *Journal of the Korean Society of Clothing & Textiles* 22 (8): 1001–1010.

Choi, Woojin, Seyoon Jang, Ha Youn Kim, Yuri Lee, Sang-goo Lee, Hanbit Lee, and Sungchan Park. 2023. "Developing an AI-Based Automated Fashion Design System: Reflecting the Work Process of Fashion Designers." *Fashion & Textiles* 10 (1): 39. https://doi.org/10.1186/s40691-023-00360-w.

Csanák, Edit. 2019. *Craft & Tradition: Challenges of Reinterpreting National Heritage in Fashion and Textile Design*. University of Novi Sad, Technical Faculty Mihajlo Pupin.

Elmourabit, Zohair, Asmaâ Retbi, and Nour-Eddine El Faddouli. Oct. 23, 2024. "The Impact of Generative Artificial Intelligence on Education: A Comparative Study." *European Conference on e-Learning* 23 (1): 470–476. https://doi.org/10.34190/ecel.23.1.2975.

Finneran, Christina M., and Ping Zhang. 2005. "Flow in Computer-Mediated Environments: Promises and Challenges." *Communications of the Association for Information Systems* 15 (1): 4. https://doi.org/10.17705/1CAIS.01504.

Guo, Ziyue, Zongyang Zhu, Yizhi Li, Shidong Cao, Hangyue Chen, and Gaoang Wang. 2023. "AI Assisted Fashion Design: A Review." *IEEE Access* 11: 88403–88415. https://doi.org/10.1109/ACCESS.2023.3306235.

Hee, Heo Yoon. 2023. 김해·고령 등 '가야고분군' 7곳, 유네스코 세계문화유산 됐다. [Seven 'Gaya Ancient Tombs' Including Gimhae and Goryeong Became UNESCO World Heritage Sites]. *Chosun*. Accessed April 27, 2024, https://www.chosun.com/culture-life/culture_general/2023/09/17/CFPPONZODZBCZJOEFMWHDMZFFI/.

Hong, Soo-Nam, and Sang-eun Lee. 2012. "A Study of Formative Aspects of Romanticism in Modern Wedding Dresses." *Journal of the Korea Fashion & Costume Design Association* 14 (3): 123–134.

Jeon, Youngseung, Seungwan Jin, Patrick C. Shih, and Kyungsik Han. 2021. "FashionQ: An Ai-Driven Creativity Support Tool for Facilitating Ideation in Fashion Design." In *Proceedings of the 2021 CHI Conference on Human Factors in Computing Systems*, 1–18. https://doi.org/10.1145/3411764.3445093.

Jonson, Ben. 2005. "Design Ideation: The Conceptual Sketch in the Digital Age." *Design Studies* 26 (6): 613–624.

Jung, Dawool, and Sungeun Suh. 2024. "Enhancing Soft Skills through Generative AI in Sustainable Fashion Textile Design Education." *Sustainability* 16 (16): 6973. https://doi.org/10.3390/su16166973.

Kim, Ji Young, and Yun Jeong Oh. Nov. 30, 2012. "'Fashion-Cultural Products Design Using Cultural Heritage – Focused on Jikji, Memory of the World of UNESCO -.' The Korean Society of Costume." *Korean Society of Costume* 62 (7): 133–149. https://doi.org/10.7233/jksc.2012.62.7.133.

Kim, Kyeng Tae. 2017. "A Study on the Patterns of Gaya Cultural Relics for Jewelry Product Development." *Design Convergence Research* 16 (1): 237–248.

Kim, Se Jin. 2024. "Generative Artificial Intelligence in Collaborative Ideation: Educational Insight from Fashion Students." *IEEE Access* 12: 49261–49274. https://doi.org/10.1109/ACCESS.2024.3382194.

Kizilcec, René F., and Hansol Lee. 2009. "Algorithmic Fairness in Education." In *The Ethics of Artificial Intelligence in Education*, Vol. 2, 174–202. Routledge. https://doi.org/10.4324/9780429329067-10.

Lee, Eun-Sook, and Eun-Jung Lee (eds.). 2009. "A Study on the Preference of Wedding Dress Design-Focused on Ulsan and Seoul." *Journal of the Korea Fashion and Costume Design Association* 11: 117–129.

Lee, Sanghee, and Sookhee Kwon. 2023. "Dress Design Analysis and Pattern Making Techniques for Increasing the Efficiency of Dress Production." *Human Ecology Research* 61 (1): 53–71. https://doi.org/10.6115/her.2023.005

Lee, Yoon Kyung. 2022. "How Complex Systems Get Engaged in Fashion Design Creation: Using Artificial Intelligence." *Thinking Skills & Creativity* 46: 101137. https://doi.org/10.1016/j.tsc.2022.101137.

Leeum Museum. n.d. 금관 *[Gold Crown]*. Accessed January 2, 2025, https://www.leeumhoam.org/leeum/search?keyword=%EA%B8%88%EA%B4%80

Liu, Linlin, Haijun Zhang, Yuzhu Ji, and Q. M. Jonathan Jonathan Wu. 2019. "Toward AI Fashion Design: An Attribute-GAN Model for Clothing Match." *Neurocomputing* 341: 156–167. https://doi.org/10.1016/j.neucom.2019.03.011.

Rostamzadeh, Negar, Seyedarian Hosseini, Thomas Boquet, Wojciech Stokowiec, Ying Zhang, Christian Jauvin, and Chris Pal. 2018. "Fashion-Gen: The Generative Fashion Dataset and Challenge." *arXiv Preprint ArXiv:1806.08317.*

Vartiainen, Henriikka, and Matti Tedre. 2023. "Using Artificial Intelligence in Craft Education: Crafting with Text-to-Image Generative Models." *Digital Creativity* 34 (1): 1–21. https://doi.org/10.1080/14626268.2023.2174557.

Xu, Q., Hubin Liu, Yulong Liu, and Shan Wu, [Retracted]. 2021. "Innovative Design of Intangible Cultural Heritage Elements in Fashion Design Based on Interactive Evolutionary Computation." *Mathematical Problems in Engineering [Online]* 2021: 1–11 Accessed December 29, 2024, https://www.hindawi.com/journals/mpe/2023/1234567/, https://doi.org/10.1155/2021/9913161.

Yan, Han, Haijun Zhang, Linlin Liu, Dongliang Zhou, Xiaofei Xu, Zhao Zhang, and Shuicheng Yan. 2023. "Toward Intelligent Design: An Ai-Based Fashion Designer Using Generative Adversarial Networks Aided by Sketch and Rendering Generators." *IEEE Transactions on Multimedia* 25: 2323–2338. https://doi.org/10.1109/TMM.2022.3146010.

BIOGRAPHY

Si Eun Kim is a beauty expert specializing in makeup artistry and currently pursuing a Ph.D. in the Interdisciplinary Program in Senior Human Ecology at the Graduate School of Changwon National University. Her research focuses on fashion marketing, with a specific emphasis on integrating beauty services into fashion marketing strategies. She is interested in how beauty services can be integrated into fashion marketing strategies to enhance consumer engagement, influence purchasing decisions, and strengthen brand connections. (https://orcid.org/0009-0002-3654-8433).

Se Jin Kim is an associate professor in the Department of Clothing and Textiles at Hanyang University. Her research areas include fashion design and fashion education methods using digital technology and changes in fashion media in the post-digital era. (https://orcid.org/0000-0002-0543-1112).

3 Reconfiguring Fashion Branding

The Role of AI in Recursive Aesthetic Evolution

Zhe Wang and Yanna Jessica Xu

3.1 INTRODUCTION, CONTEXT, AND LITERATURE REVIEW

Over the past few years, fashion brands have increasingly embraced digitalisation and virtual world-building, initially focusing on metaverse-powered innovations before shifting towards AI-driven branding strategies. The metaverse, once hailed as the future of digital interaction, fuelled the rise of virtual influencers and avatars, with brands investing in hyper-realistic digital personalities to engage online audiences. One of the most notable examples was Ayayi, China's first ultra-realistic "metahuman" influencer, launched in 2021 by Raimai Technology. Initially a sensation, Ayayi amassed a significant following on platforms like Red Note (also known as Xiaohongshu), engaging in high-profile brand collaborations. However, her prominence declined alongside the waning enthusiasm for the metaverse, as brands and consumers gravitated towards more authentic and AI-driven marketing tools (Influencer China Agency, 2023).

This transition reflects the fashion industry's evolving focus on AI-generated creativity rather than metaverse-dependent engagement. Brands are now leveraging generative AI tools such as MidJourney and Stable Diffusion, shifting from static digital influencers to dynamic AI-assisted design and marketing automation. Balmain, a pioneer in blending fashion with emerging technologies, initially explored virtual models with its "Balmain Army" but has since integrated AI into broader branding strategies (Jing Daily, 2023). Instead of relying on fully digital personas, brands are now using AI to enhance real-world campaigns, personalise customer interactions, and streamline creative production (Jones, 2022; Jing Daily, 2024). This movement from metaverse-driven digitalisation towards AI-enhanced creativity underscores a critical industry shift: from investing in virtual spaces to harnessing AI's ability to generate hyper-personalised, scalable, and efficient branding content.

The shift is apparent: instead of relying on static virtual influencers, brands are now harnessing AI's generative capabilities – tools like MidJourney and Stable Diffusion – to automate creative production, personalise customer interactions, and scale marketing efforts efficiently (Jing Daily, 2024), and they began sweeping across society with an unstoppable force, much like the impact of the spinning jenny during

 DOI: 10.1201/9781003594482-4

the First Industrial Revolution. The emergence of these practical AI tools signals the beginning of a new technological era poised to revolutionise production methodologies (Sayem, 2023). The rapid adoption of digital technologies, especially the establishment of AI and its image production applications, are significantly challenging the methods of human-directed creative production. As one of the core sectors of the creative industry, this rising trend of AI has manifested and influenced the fashion industry in production, presentation, and communication.

As the year 2025 draws, it is evident that AI tools such as MidJourney and Stable Diffusion for image generation, Runway for video generation, as well as ChatGPT and DeepSeek, for both text and creative generation, are significantly revolutionising creative production, impacting various artistic and creative processes by enhancing efficiency, accessibility, and innovation. **MidJourney** generates detailed artwork from textual prompts, enabling swift and democratised creation of complex visual art (Data Science Dojo, 2023; Nadeem, 2023). **Stable Diffusion** uses diffusion models to produce high-quality images from text descriptions, facilitating rapid prototyping and detailed visualisation (NVIDIA, 2023; KDnuggets, 2024). **ChatGPT** has evolved to include image generation capabilities, transforming detailed textual prompts into vivid visual representations (OpenAI, 2024).

AI-generated posters in subway stations, the growing presence of virtual models in advertisements, and AI-synthesised stage effects have become increasingly commonplace (Afifah, 2024). This innovative technology has proven to be a formidable ally for fashion creatives, enabling advancements from the rise of virtual influencers to the fusion of products with virtual models in advertisements. AI is effectively navigating the blurred boundary between the virtual and the real, inaugurating a new era of artistic co-creation between technology and humanity. This development demonstrates how novel creative methods can influence the presentation of fashion, offering numerous fantastical and imaginative possibilities (Kohli et al., 2023).

Existing studies, when mentioning AI and fashion presentation, have shed light on the following various perspectives. The integration of artificial intelligence (AI) within the fashion industry has heralded a significant transformation in various domains, encompassing design (Jeon et al., 2021; Guo et al., 2023; Zhang and Liu, 2024), prototyping (Huang et al., 2024), marketing, and supply chain management (Rathore, 2017a, 2017b; Luce, 2018; Oosterom et al., 2023; Afifah, 2024). AI technologies such as machine learning, decision-support systems, and optimisation have increasingly mediated fashion design and production workflows. Early experiments like Google's "Project Muze" (Rietze, 2016) used user preference data and aesthetic parameters to algorithmically generate garment proposals, while contemporary luxury brands have applied AI systems in digital look generation, show scenography, and creative prototyping (Epic Games, 2021). Furthermore, AI enhances trend forecasting and supply chain efficiencies, aiding fashion companies in making informed decisions and offering personalised shopping experiences through virtual and augmented reality. AI's role in defect detection and sales forecasting has also contributed to more sustainable manufacturing practices by reducing waste (Rathore, 2017b; Candeloro, 2020; Pillarisetty and Mishra, 2024).

Oosterom et al. (2023) discuss the growing presence of digital humans and the metaverse in fashion marketing. Virtual influencers and avatars, created using advanced AI and CGI technologies, are becoming integral to brand promotion and consumer engagement. These digital entities bridge the gap between the virtual and real worlds, offering new avenues for creative expression and consumer interaction in the fashion industry (Kohli et al., 2023). This trend is expected to expand beyond social media into various facets of digital interaction, suggesting a significant shift in how brands may engage with consumers in the future.

However, the extant research indicates that earlier AI innovations in the fashion industry were predominantly centred on garment pattern design and consumer experience enhancement at commercial endpoints. In previous phases before mid-2022, AI had yet to democratise creative production methods or harness the advanced image generation capabilities emblematic of the Web 3.0 era. Consequently, although there were speculative discussions on the future impact of virtual idols and social media on branding strategies, the absence of these advanced AI tools precluded any substantial exploration of this domain. Thus, studies published before 2024 did not address the transformative potential of post-2022 AI tools such as MidJourney, Stable Diffusion, and ChatGPT, whose advancements have since revolutionised image generation, fundamentally reshaping digital fashion innovation and branding strategies.

Furthermore, it can be witnessed that the barriers to creative production in the fashion industry have significantly diminished, leading to profound shifts in the conception and execution of creative campaigns since the introduction of AI tools in mid–2022. These changes hold the potential to fundamentally redefine the systems, methodologies, and logics of presenting and disseminating creativity within the fashion sector. This transformation is already unfolding subtly in China: Percy Lau transitioned from eyewear design to operating DESIGNATOR AI for customised models; Xu Xiaowei integrated AI into photography, expanding creative potential while highlighting the irreplaceability of human insight; Macci Leung adopted AI for innovative visual content, showcasing its role in augmenting human creativity; and Zhou Baibai revitalised her creative enthusiasm through AI, gaining visibility in prestigious publications. These examples evidenced AI's profound impact, fostering innovation and redefining traditional practices while preserving the human element, posing new questions about fashion production and communication.

Despite the profound implications of these developments, research in this area remains sparse. Our study seeks to address this gap by closely tracking industry changes and providing a thorough analysis of AI's impact on fashion branding. In particular, it aims to answer the question: How are AI-driven innovations reshaping creative production and branding strategies in the fashion industry, and what implications do these changes have for the future of fashion production and communication?

This research will offer pioneering insights and theoretical frameworks at the nexus of AI and fashion branding. It will chart new directions for creative production in fashion and elucidate how AI-driven innovations can reshape branding strategies. By exploring this emerging field, our study will contribute significantly to both academic discourse and practical applications, positioning itself at the forefront of technological and creative advancements in the fashion industry.

3.1.1 Research Question and Aim

This research aims to investigate how are AI-driven innovations reshaping creative production and branding strategies in the fashion industry, and what implications do these changes have for the future of fashion production and communication, using Percy Lau's DESIGNATOR AI studio and its collective works as a case study? Demographically, the central question addresses the specific roles AI plays in the branding efforts of this unique studio and its influence on the fashion industry in China. Furthermore, this analysis will introduce a paradigm for future studies on how image-generation AI tools will revolutionise creative production and communication globally, beyond China.

3.1.2 Objectives

1. To explore the application of AI in the branding process at DESIGNATOR AI, including its impact on design methodology and fashion production (advertising in particular).
2. To assess how AI-driven branding techniques can sustain and evolve a fashion brand's identity and aesthetic values.
3. To identify the implications of AI technology for future fashion branding strategies.

3.2 METHODS

3.2.1 Case Study Selection: Percy Lau and DESIGNATOR AI

This study employs a case study with qualitative interview and observation as main methods to explore how AI-driven innovations are reshaping creative production and branding strategies in the fashion industry. Percy Lau and her studio, DESIGNATOR AI, are selected for their innovative integration of advanced AI models, such as Stability AI's stable diffusion API, to create custom fashion designs. Her pioneering methodologies for training AI models to understand and replicate fashion aesthetics, as seen in collaborations with designers like Xander Zhou and Chen Peng, highlight the studio's role in generating both replicative and innovative designs. Additionally, DESIGNATOR AI's comprehensive services, from digital models to editorial projects, demonstrate practical AI applications across various facets of fashion production, making it an exemplary subject for studying the future implications of AI in fashion.

3.2.1.1 An Overview of Percy Lau and the AI Operations of DESIGNATOR AI Studio

Percy Lau (LAU HOI TSE) is a distinguished eyewear accessory designer and digital artist recognised for his avant-garde and experimental designs. A graduate of Central Saint Martins College of Art and Design, Lau received the International Talent Support YKK Award in 2013, subsequently founding his eponymous brand, PERCY LAU. Her work is acclaimed for pushing the boundaries of contemporary eyewear design, utilising innovative technologies such as Shapr3D to enhance design efficiency (Shapr3D Blog, 2023). *The Optical Journal* describes his eyewear as "spectacles in their own right" (*The Optical Journal*, 2015), highlighting the evolution from humble beginnings to highly coveted pieces. Byrdie commends her brand for its futuristic edge and sleek silhouettes, placing it among the best sunglasses brands for style and shade (Byrdie, 2022).

Lau's eyewear has been featured in over 50 fashion magazines globally and worn by more than 100 artists worldwide. Her contributions to the industry have been acknowledged through his inclusion in the Vogue Talent list and Forbes 30 Under 30 in both China and Asia (Luxe.co, 2020). In 2020, she launched the first domestic collection of virtual fashion accessories, demonstrating her forward-thinking approach. Lau continues to influence the fashion world with her cutting-edge designs and commitment to redefining eyewear norms. In 2023, she founded DESIGNATOR AI Studio, providing bespoke modelling services and workflow optimisations for multiple brands, further solidifying her position as an industry leader.

Following the indefinite suspension of her eponymous eyewear brand in May 2023, Percy Lau promptly ventured into a new enterprise – the creative studio DESIGNATOR AI. Through this studio, which offers customised AI data models to both private individuals and businesses, Lau has gained unprecedented insights into the role of AI as a powerful assistant in the creative process. Contrary to prevalent fears that AI might supplant human jobs, her extensive experience with AI has only reinforced her belief in the indispensable nature of human creativity. The studio focuses on three primary areas: digital models, customised design through model training, and project-based outputs including editorials and product releases.

3.2.1.1.1 Digital Models

One of the key innovations at DESIGNATOR AI is the creation of digital models. It primarily provides personalised AI models (such as Control Net, DreamBooth/LORA) for individuals and businesses, based on the Stable Diffusion programme – a type of AI image generation tool. These models are designed using AI algorithms that incorporate various design elements and aesthetic principles. By training these models on vast datasets, the studio can generate highly realistic and diverse digital representations of fashion items. For example, if a designer needs a custom design model, they can upload images from their previous collections into the model's database. The AI can then generate a new series that aligns with the designer's aesthetic style and maintains continuity with previous collections, all in a fraction of the time it would typically take. According to Lau (2024), the digital models are primarily used to create fashion items that resonate with contemporary trends and aesthetics. The use of Stability AI's stable diffusion API allows the studio to refine these models further, integrating open-source tools to enhance creativity and customisation (Lau, 2024).

The digital models (see Figure 3.1) produced by DESIGNATOR AI are not just replicas of existing fashion items but are designed to push the boundaries of creativity. By integrating AI with traditional design processes, the studio can experiment with new forms and styles that might be difficult or time-consuming to achieve manually. This capability allows DESIGNATOR AI to stay ahead of fashion trends and offer unique products to its customers.

FIGURE 3.1 Digital model created based on the concept of "AI making skins for Y2K aliens". (Courtesy to Percy Lau, Copyright@DESIGNATOR)

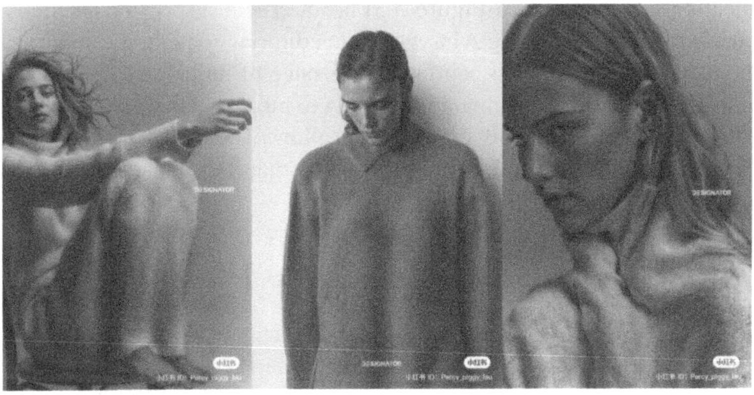

FIGURE 3.2 Input of French female model wearing a sweater; example renderings generated for a brand client.

(Courtesy to Percy Lau, Copyright@DESIGNATOR)

3.2.1.1.2 Customised Fashion Design

The studio also offers customised design services, leveraging AI to create tailored designs for individual clients. This involves training AI models on specific design languages and aesthetic preferences, allowing for the production of unique fashion items that align with the client's vision. Lau notes that this customisation is achieved through a detailed understanding of design logic and language, which AI can learn and replicate effectively (Lau, 2024). By using Control Net, DreamBooth and Lora models, DESIGNATOR AI can produce designs that maintain high fidelity to the client's desired aesthetic while introducing innovative elements (Lau, 2024).

Customised design (see Figure 3.2) is a significant part of DESIGNATOR AI's business model. By offering personalised fashion items, the studio can cater to a niche market that values exclusivity and individualism. The use of AI in this context not only enhances the efficiency of the design process but also ensures that each product is unique and tailored to the client's preferences. This approach has helped

DESIGNATOR AI build a loyal customer base and establish itself as a leader in AI-driven fashion design.

3.2.1.1.3 Project-Based Outputs

DESIGNATOR AI produces project-based outputs such as editorials and product releases. These projects utilise AI to streamline the design process, from concept development to final production. The use of AI not only enhances the efficiency of these processes but also introduces new creative possibilities that were previously unattainable. For example, the studio's use of AI to generate unique editorial content has been instrumental in maintaining a fresh and engaging brand presence (Lau, 2024). The AI-generated content is often indistinguishable from human-created designs, showcasing the potential of AI to augment human creativity in fashion (Lau, 2024).

Project-based outputs (see Figure 3.3) are a crucial aspect of DESIGNATOR AI's operations. By leveraging AI to generate editorial content, the studio can produce high-quality visuals that capture the essence of its brand. These visuals are used in various marketing campaigns, helping to build and maintain a strong brand identity. The use of AI in this context also allows DESIGNATOR AI to experiment with different styles and formats, ensuring that its content remains fresh and relevant.

3.2.2 Data Collection: In-depth Interview and Visual Experimentation

Data for this case study were collected through a series of in-depth interviews (five and a half hours in total) with Percy Lau and an analysis of DESIGNATOR AI's AI-generated outputs and design processes. We also undertook a series of visual experimentation, asked Percy to show us how she utilises stable diffusion as a tool to train style models, generating the visual presentations required by client brands through the input of prompts. This process encompasses creating runway lookbooks, advertising campaigns, and conducting style mixing experiments. For instance, Percy has shown how she can seamlessly integrate the styles of various deceased photographers into the brand's DNA, ensuring that these integrations are achieved without infringing on copyrights. Through this approach, Percy effectively maintains the core identity of the client's brand while innovatively enhancing its visual appeal. The specific experimentations we undertook during the investigation are listed as follows:

3.2.2.1 Runway Lookbook

1. **Xander Model Experiment**:
   - Percy initially used early Xander models for training, which had limited information. However, by increasing the data volume, the model's precision improved, allowing it to learn more intricate design details. This experiment demonstrated the potential of AI to capture and replicate the nuanced aesthetics of a designer's work. See the following output in Figure 3.4.

FIGURE 3.3 A fashion photography branding campaign created for the client FLEX VENUS using an AI model.

(Courtesy to Percy Lau, Copyright@DESIGNATOR)

2. **Chen Peng Runway Design Training**:
    - Percy, with Chen Peng's authorisation, trained the model on Chen Peng's runway designs. These designs, although from previous collections, provided new season design inspiration through AI-generated images. This experiment showed how AI could rejuvenate and innovate within established design frameworks. See the following output in Figure 3.5.

FIGURE 3.4 Output of Xander Model Experiment.

(Courtesy to Percy Lau, Copyright@DESIGNATOR)

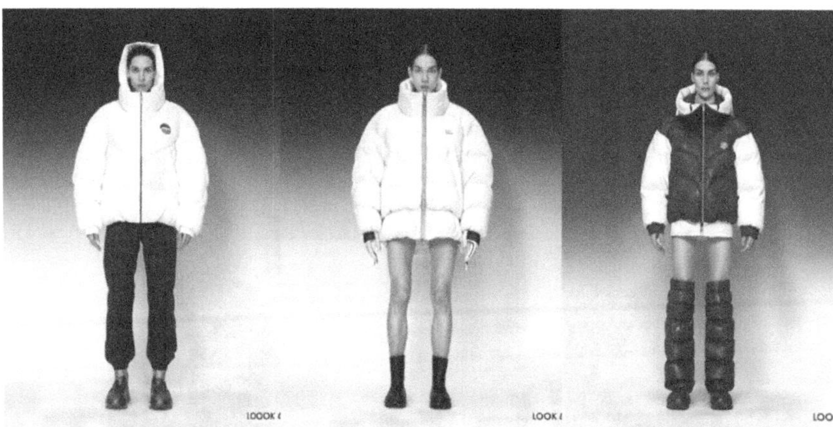

FIGURE 3.5 Output of Chen Peng Runway Design Training.

(Courtesy to Percy Lau, Copyright@DESIGNATOR)

3.2.2.2 Advertising Campaign

1. **Model Detail Optimisation Experiment**:
    - Percy generated model images using AI and refined details based on client feedback. Despite AI generating the background, characters, and makeup, product details were still photographed to ensure quality. This approach ensured the seamless integration of AI and traditional photography to maintain high standards in advertising visuals.. See the following output in Figure 3.6.
2. **Brand Advertising Creativity Experiment**:
    - In creating advertising materials, Percy combined different elements, such as flowers, with the model to create unique and surprising visual

FIGURE 3.6 Output of Model Detail Optimisation Experiment.

(Courtesy to Percy Lau, Copyright@DESIGNATOR)

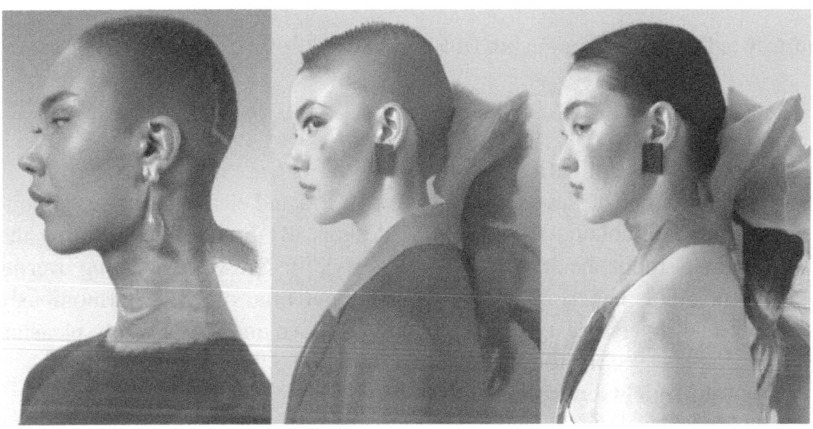

FIGURE 3.7 Output of Brand Advertising Creativity Experiment.

(Courtesy to Percy Lau, Copyright@DESIGNATOR)

effects. This method enhanced the creativity of the brand's advertisements, demonstrating AI's capability to produce innovative and engaging content. See the following output in Figure 3.7.

3.2.2.3 Style Mixing Experiments

1. **Xander and Chen Peng Style Mixing Experiment**:
   - Percy experimented with mixing the styles of Xander and Chen Peng by controlling different model ratios. He extracted Xander's silhouettes and colours and combined them with Chen Peng's materials. This careful balancing avoided overfitting and resulted in designs that effectively merged the distinct styles of both designers. See the following output in Figure 3.8.

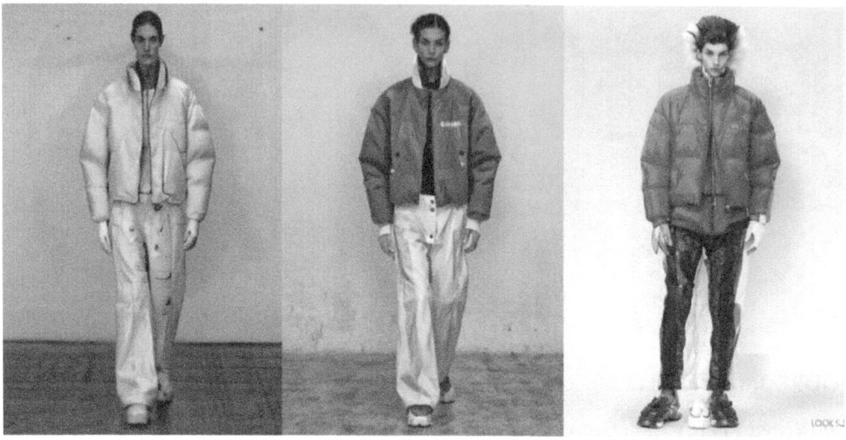

FIGURE 3.8 Output of Xander and Chen Peng Style Mixing Experiment.

(Courtesy to Percy Lau, Copyright@DESIGNATOR)

2. **Control Net Application Experiment**:
   - Using Control Net technology, Percy extracted elements such as silhouettes and colours from multiple designs. By precisely controlling the model ratios, he achieved seamless style fusion while avoiding overfitting. This experiment showcased the potential of AI to harmoniously blend diverse design elements into cohesive and aesthetically pleasing outputs. See the following output in Figure 3.9.
3. **Digital Model Generation**:
   - Percy developed digital models and avatars, which were used for various projects. This included editorial content and product generation. The digital models were customised to reflect the brand's identity and aesthetic, demonstrating the versatility of AI in creating personalised and brand-consistent digital assets. See the following output in Figure 3.10.

These visual experiments collectively showcased Percy Lau's innovative use of AI in fashion design, pushing the boundaries of traditional design methods and creating new possibilities for visual presentation. By continuously adjusting and refining AI models, Percy successfully merged different design elements, ensuring that the outputs remained true to the brand's identity while being rich in creativity.

The interviews provided detailed insights into the specific AI technologies used, the creative workflows adopted, and strategic decisions guiding AI integration in fashion production. Conversations covered the development and training of AI models, the creation of digital fashion content, and the challenges and opportunities presented by AI. This qualitative data was meticulously recorded and analysed to extract key themes and insights relevant to the research objectives. Additionally, a technical analysis assessed the quality, creativity, and industry applicability of AI-generated

FIGURE 3.9 Left Image: Utilising minimalism to blend the work of artist Rothko, without adding prompt words for a fashion model.

Right Image: Utilising minimalism to blend the work of artist Rothko, adding prompt words for a fashion model.

(Courtesy to Percy Lau, Copyright@DESIGNATOR)

FIGURE 3.10 Output of Digital Model Generation Experiment.

(Courtesy to Percy Lau, Copyright@DESIGNATOR)

designs, while a comparative evaluation juxtaposed AI-generated and traditional designs to highlight unique contributions and divergences in creative processes.

3.2.3 Data Analysis Method and Theoretical Framework

Discourse analysis is employed as the primary method for analysing the interview data. This method involves identifying recurring themes and patterns in the transcripts, coding significant concepts and terminologies used by Percy Lau, and examining the broader context of the fashion industry and AI innovations. The analysis focuses on language and rhetorical strategies to articulate AI capabilities and limitations, her vision for the future of fashion, and her responses to industry challenges. By considering intertextuality and interdiscursivity, the study analyses how Percy Lau's discourse interacts with other discourses within fashion and technology industries. This comprehensive approach provides nuanced insights into the impact of AI on fashion production and branding strategies.

The primary theoretical framework applied in this discussion is "recursive modernity" (Giddens, 1991; Beck, 1992; Ulrich and Giddens, 1994; Lagerkvist, 2013; Sayre, 2013; Bratton, 2016; Fischer, 2018; Ishizumi, 2022), which provides a critical foundation for analysing AI's role in revolutionising fashion branding, particularly in the context of data-driven creative production. The concept of recursive modernity has emerged as a critical framework for understanding the non-linear, feedback-driven nature of contemporary social, political, and technological transformations. While not universally defined by a single theorist, the term has been employed across various disciplines, including political theory (Ulrich and Giddens, 1994; Ishizumi, 2022), urban studies (Lagerkvist, 2013; Sayre, 2013), science and technology studies (STS) (Bratton, 2016; Fischer, 2018), and media analysis (Sayre, 2013; Bender, 2024). The fundamental premise of recursive modernity is that modernisation does not progress in a linear trajectory but continuously loops back on itself, incorporating historical feedback, structural adaptation, and iterative recalibration.

Early discussions of recursion in modernity can be traced to Anthony Giddens (1991) and Ulrich Beck (1992), who introduced reflexive modernity as a framework for understanding how modern societies constantly reassess and reconfigure themselves. Giddens (1991) argues that modern institutions are in a state of continuous transformation, informed by recursive feedback mechanisms that challenge traditional notions of stability and progress. Similarly, Beck (1992) contends that modernity generates its own risks and crises, necessitating a perpetual process of revision and adaptation. More recent engagements with recursive modernity have expanded beyond sociology into governance, urbanism, and technology studies. Sayre (2013) explores how disaster preparedness in Japan operates within a recursive modernity framework, where policies and security measures are constantly restructured based on historical events and evolving risk assessments. Similarly, Lagerkvist (2013) applies the concept to Shanghai's urban transformation, demonstrating how the city's architectural and cultural shifts reject immediate pasts while cyclically reviving older historical aesthetics. This exemplifies a recursive negotiation with historical identity, where modernity does not abandon its origins but instead reinterprets them in new forms. In the field of science and technology studies (STS), Fischer (2018) positions recursive modernity as a defining feature of contemporary governance and computational infrastructures, where AI, platform economies, and algorithmic governance create self-regulating, feedback-driven systems. This perspective aligns with Bratton's (2016) conceptualisation of planetary-scale computation, which, while not explicitly referencing recursive modernity, illustrates how technological modernity operates in modular, self-adjusting cycles. Furthermore, Ishizumi (2022) explicitly defines recursive modernity as a dialectical process within political modernisation, arguing that modernity remains open to revision and critique rather than progressing along a fixed ideological path. His work reinforces the notion that modern political and economic structures are constantly reconfigured in response to historical cycles and shifting global contexts.

Unlike traditional models of technological progress that emphasise radical innovation, the application of recursive modernity provides an insightful framework for analysing how AI revolutionises fashion branding practices. AI-driven branding does not introduce a radical departure from traditional brand identity, but rather operates in a recursive feedback loop, where historical brand aesthetics, consumer data, and algorithmic decision-making constantly refine and reshape marketing strategies. This perspective is particularly relevant for examining how AI-generated branding, as seen in Percy Lau's DESIGNATOR AI Studio, does not replace human creativity but instead operates as an iterative system that synthesises past and present design elements. This also aligns with recursive modernity's core premise that **modern** branding evolves through iterative adjustments rather than through singular, disruptive shifts.

By applying recursive modernity, this study justifies the use of AI-generated outputs as legitimate branding data. AI-driven models such as Stable Diffusion, using Control Net (an extension of Stable Diffusion that provides greater control over image structure), and DreamBooth (fine-tunes Stable Diffusion models for personalised image generation) as learned from historical and contemporary visual data, ensuring stylistic coherence while enabling scalability in fashion marketing. Therefore, recursive modernity provides a theoretical basis for understanding AI's transformative yet continuity-preserving role in fashion branding, demonstrating how fashion branding practices evolve recursively rather than disruptively.

3.3 DISCUSSION AND ANALYSIS

This section critically examines the profound impact of AI across key areas, including fashion design, production, digital heritage, and branding, highlighting both its transformative potential and the challenges it presents. The discussion explores how AI is augmenting creative practices, reconfiguring traditional labour structures, and altering the dynamics of brand communication, providing a comprehensive analysis of its implications for the future of the fashion industry, as well as theoretical studies on fashion branding.

3.3.1 Impact on Fashion Design and Production: Recursive Modernity and AI in Visual Experimentation

The visual experiments undertaken in this project, particularly in AI-driven fashion and visual design, demonstrate how computational tools are reshaping creative processes. Rather than merely accelerating design workflows, AI actively participates in aesthetic interpretation and iterative refinement. Notable examples include Lau's exploratory projects, such as the Xander Model Experiment and the Chen Peng Runway Design Training, as well as AI-assisted reinterpretations of Mark Rothko's artistic style. These cases illustrate AI's potential to absorb, recombine, and refine existing design languages, while also revealing the system's inherent constraints.

The Xander Model Experiment exemplifies how AI learns to capture intricate design details through iterative refinement. Initially trained on a limited dataset, early versions of the model struggled with precision. However, as additional data was introduced, the AI enhanced its ability to replicate nuanced stylistic elements. This process suggests that AI does not create radical artistic ruptures but rather improves its ability to process and refine existing aesthetic codes.

Similarly, the Chen Peng Runway Design Training highlights AI's dual function in preserving and innovating within an established visual identity. With the designer's authorisation, Lau trained an AI model on past runway collections, allowing AI-generated imagery to serve as conceptual references for new designs. The resulting outputs retained the distinctive characteristics of Chen Peng's previous work while simultaneously exploring new stylistic possibilities.

A more unexpected example emerged in our visual experimentation with Mark Rothko's artistic principles. Using Stable Diffusion, we sought to reinterpret Rothko's work by integrating his abstract colour fields into AI-generated portraits. Notably, despite no explicit emotional prompts, the resulting images conveyed solitude and melancholy – qualities often associated with Rothko's paintings. This suggests that AI does not merely replicate form but may also engage in a latent process of emotional translation, interpreting and synthesising affective cues embedded within its dataset.

The creative processes observed in these AI experiments align with the concept of recursive modernity, which challenges the notion of linear artistic progress. Rather than being defined by successive breaks with the past, modernity – especially within the computational era – is shaped by continuous, feedback-driven reformulations.

It has been acknowledged that modernity does not progress through radical aesthetic ruptures but rather through continuous recombination and reconfiguration of historical forms (Kellner, 1988; Huyssen, 2003; Friedman, 2015; Tomba, 2019). AI tools such as Stable Diffusion, as observed in these fashion and art experiments, does not introduce completely novel aesthetic languages but instead operates within a recursive system of adaptation, where existing styles are reprocessed into new visual configurations. By "new", we refer not simply to novel recombination but to deep stylistic evolutions, such as the transition from Baroque to Rococo or from Art Deco to Bauhaus. These shifts represent more than stylistic modifications; they constitute recursive ruptures, where modernity actively reconfigures itself rather than merely iterating upon past forms. At present, AI is largely unable to achieve such transformations independently, as its generative process is constrained by the parameters of its training data rather than by a capacity for conceptual reinvention.

From a design perspective, AI's increasing role in the visual experimentation phase exemplifies the recursive nature of contemporary creative processes.

Rather than merely accelerating pre-existing workflows, AI operates within a feedback-driven system, continuously iterating upon past visual data. This recursion – an inherent feature of AI's computational structure – enhances efficiency and facilitates a dynamic interplay between established aesthetic conventions and emerging forms. The fundamental mechanism underpinning AI's creative function lies in its ability to extract and recombine accessible visual elements from its database, responding to prompts and designer input to generate diverse permutations. In doing so, AI expands the field of potential outcomes within significantly shorter timeframes, offering creators an extensive range of possibilities for exploration.

However, within the framework of recursive modernity, AI's limitations become particularly apparent when compared to human cognition. AI's reliance on recombination – while powerful in generating variations – ultimately operates within the confines of its encoded datasets. Unlike human designers, who engage in radical stylistic departures that redefine aesthetic paradigms, AI remains primarily an iterative mechanism rather than an originator of fundamental artistic shifts.

Nevertheless, our visual experimentation with Lau produced one of the most unexpected and conceptually significant outcomes in the reinterpretation of Rothko's work (see Figure 9). In this instance, Stable Diffusion exhibited a degree of creativity that appeared to exceed its conventional recombinatory function, suggesting a more complex interaction with visual and emotional archives. Notably, despite no explicit instructions regarding emotional expression, the resulting images conveyed an affective depth often associated with contemporary art. The AI-generated portraits and facial expressions reflected a sense of solitude and melancholy, indicating a latent ability to synthesise and reinterpret emotional intensity within the visual dataset (Manovich, 2020).

This outcome demonstrates that AI does not merely reproduce or replicate, but rather reframes modernity recursively by dynamically responding to its input conditions. In this case, Stable Diffusion appears to have internalised the emotive properties of Rothko's oeuvre – not simply as formal visual data, but as an embedded aesthetic logic within its generative framework. Consequently, when tasked with incorporating Rothko's style into human forms and expressions, the AI produced outputs that were not only stylistically consistent but also evocative of the affective and existential depth that characterises Rothko's artistic legacy.

While this does not equate to a truly autonomous reinterpretation in the human sense, it suggests AI's capacity for recursive interpretation – whereby historical aesthetic principles are continuously reprocessed, recontextualised, and re-expressed within new visual environments. This raises compelling questions regarding AI's evolving role in creative practice: Will AI remain a recursively modern tool, iterating within established frameworks? Or will future developments enable a deeper engagement with the ruptural dimensions of artistic evolution, where entirely new visual paradigms emerge beyond human-designed stylistic precedents?

At present, AI functions predominantly within the recursive modernity framework, iterating upon historical aesthetics rather than producing fundamentally new artistic paradigms. The fashion and art experiments discussed earlier demonstrate that AI's role is primarily one of adaptive synthesis, reconfiguring past stylistic elements into novel yet recognisable forms.

3.3.2 Expansion in Digital Heritage: Reconfiguration, Not Replication

In the context of digital heritage, recursive modernity challenges the notion that cultural preservation operates through static archival methods. Instead, it suggests that heritage is not merely conserved but continuously reshaped through computational infrastructures, iterative reinterpretation, and digital augmentation. AI tools such as Stable Diffusion and Control Net exemplify this shift, where heritage is no longer a fixed historical reference but a fluid and adaptive aesthetic system that evolves within computational frameworks.

Lau's AI-driven experiments provide concrete examples of how digital heritage operates within recursive modernity. In her Xander and Chen Peng style fusion experiment, AI was trained to combine historical silhouettes and material applications, not by simply layering past designs but through an iterative refinement of encoded aesthetic rules (Lau, 2024). This process aligns with recursive modernity's premise that historical forms are not static artifacts but evolving datasets, continuously informing and transforming contemporary creative expressions.

Unlike traditional archival preservation, which aims to maintain historical integrity, AI-driven processes reinterpret heritage through continuous recombination. The application of Control Net as an extension of Stable Diffusion in fashion design demonstrates how AI can extract, recombine, and reconstruct past aesthetics without simply reproducing them (Lau, 2024). This reflects recursive modernity's principle that digital tools do not merely store the past but actively reshape it through algorithmic engagement. For example, in Lau's digital model training for campaign visuals, AI-generated avatars were not designed to replace human subjects but to simulate, extend, and recursively refine brand aesthetics. These digital avatars, trained on brand-specific datasets, allowed for a consistent yet flexible evolution of brand identity, ensuring that heritage aesthetics remained fluid rather than fixed (Lau, 2024).

Therefore, AI serves not as a passive digital archive but as a recursive engine that continuously remixes past aesthetics into new configurations. This challenges conventional ideas of historical authenticity, suggesting that heritage is not a static entity to be preserved but an evolving system to be engaged with dynamically. The impact of recursive modernity on digital heritage challenges traditional preservationist models, suggesting that heritage is not a static legacy to be safeguarded but an evolving framework to be continually engaged with. AI does not merely document the past – it actively transforms it, generating new forms of engagement with historical aesthetics through continuous computational remixing.

While AI expands creative possibilities, it simultaneously raises critical ethical, legal, and economic concerns. Ensuring that AI serves as a mechanism for thoughtful cultural engagement rather than passive replication will determine whether digital heritage remains a recursively modern system of evolution or a site of unchecked commodification.

3.3.3 Impact on Fashion Production: Redefining Production Costs, Creative Roles, and Labour Dynamics

The integration of AI technologies into fashion production is dramatically lowering costs, making high-quality creative production more accessible to SMEs (small and medium-sized enterprises). Lau (2024) highlights that AI-generated fashion campaigns cost only "one-fifth of traditional productions that rely on human teams, significantly reducing labour, studio rental, and post-production expenses". This cost efficiency enables smaller brands to experiment with high-quality visuals, expanding creative possibilities that were previously restricted by budget constraints.

For many SMEs, the ability to use AI-generated models, backdrops, and styling removes barriers to entry in fashion marketing. Previously, producing a campaign with high-end aesthetics required substantial investment in photographers, makeup artists, stylists, models, and set designers. AI, however, offers a scalable alternative, enabling brands to rapidly create campaign imagery without the logistical complexities of a physical shoot. Lau (2024) observes that this democratisation of creative production benefits emerging designers and independent brands, "offering greater flexibility and experimentation in visual storytelling". The reduced cost structure allows designers to test multiple campaign aesthetics, iterate quickly, and refine brand narratives without being constrained by traditional production budgets.

While AI-driven cost reduction benefits creative accessibility, it also raises critical concerns about labour displacement in the fashion industry (Bender, 2024; Erickson, 2024). The decline in demand for traditional production teams, including photographers, stylists, and set designers, raises questions about the future role of human labour in creative industries. Lau (2024) acknowledges this concern, noting that while AI can replicate technical execution, it struggles to capture the tactile richness and organic spontaneity of human-led creative direction. This echoes the broader debate on AI's impact on craftsmanship, where the automation of image generation and branding workflows could lead to a shift in labour demand rather than complete redundancy (Bender, 2024; Erickson, 2024).

The challenge, then, is not merely whether AI replaces traditional creative jobs but rather how the industry adapts to this shift. As AI-powered workflows become standard practice, brands will need to develop hybrid models where AI augments rather than replaces human expertise. Lau (2024) highlights that AI does not eliminate the need for creative direction; rather, it requires a different skill set, where designers curate datasets, refine AI outputs, and maintain aesthetic consistency across campaigns. Instead of replacing jobs outright, AI is shifting the nature of creative expertise towards a combination of artistic sensibility and computational literacy.

AI's role in reducing production costs is a double-edged sword. On one hand, it lowers barriers for SMEs, making high-quality campaigns financially feasible for independent designers and emerging brands. On the other hand, it raises unresolved debates on labour displacement, requiring the fashion industry to redefine the relationship between AI and human craftsmanship. The future of AI in fashion production will depend on how brands balance efficiency with ethical responsibility. As AI becomes a standard tool in creative workflows, the industry must ensure human talent remains central, with AI serving as an enhancer rather than a replacement. Whether AI is a tool for democratising creativity or a catalyst for job displacement will be determined by how brands, policymakers, and creatives negotiate this transition. What is clear, however, is that AI is no longer just an auxiliary tool – it is actively reshaping the economic and creative landscape of the fashion industry.

3.3.4 Impact on Fashion Branding and Communication: Transformation of Creative Labour and Algorithmic Identity

The transformation of branding and communication practices through AI-driven content generation aligns with the principles of recursive modernity, wherein aesthetic identities are continuously reprocessed rather than existing as fixed cultural markers. This recursive dynamic not only restructures creative labour in fashion branding but also raises critical questions about the evolving role of human agency in an AI-mediated industry. Bender (2024) examines this shift in the context of Generative AI's impact on media industries, arguing that AI does not simply replace human creativity but fundamentally reconfigures creative labour, altering the mechanisms of artistic production and decision-making. Rather than eliminating human input, AI repositions creative practitioners as curators of computationally generated outputs, introducing a new paradigm of algorithmic branding.

As branding increasingly incorporates AI-driven generative design, the standardisation of brand identity becomes an algorithmic process rather than an organic creative evolution. AI's influence has introduced a high-efficiency, data-driven approach to creativity, where brands prioritise consistency, predictability, and algorithmic refinement over the spontaneity and disruption inherent to human-led creative processes (Bender, 2024; Erickson, 2024). This closely aligns with the notion of recursive modernity, in which AI does not forge new creative directions but continuously restructures and iterates upon established aesthetic frameworks. In this context, branding ceases to be a domain of radical artistic innovation and instead becomes a process of algorithmic refinement, with human creatives acting as strategic editors rather than originators.

Lau's AI branding experiments further illustrate this recursively modern logic. In her AI-assisted branding campaigns, AI-driven models were observed to default to standardised aesthetic conventions unless actively directed otherwise (Lau, 2024). Bender (2024) highlights a similar phenomenon in AI-generated media, emphasising that Generative AI is inherently predisposed to reinforcing pre-existing creative norms rather than challenging them. This reflects the broader industrial shift towards computational branding strategies, where algorithmic efficiency is privileged over subjective artistic interpretation. AI's ability to generate high volumes of branding content at scale offers significant benefits in terms of cost and speed, yet also introduces a risk of creative homogenisation, where branding aesthetics become increasingly data-driven rather than artistically exploratory.

Thus, while AI offers scalability and operational efficiency, it also instils a recursive aesthetic cycle within branding, wherein creative outputs are continuously generated within pre-existing computational parameters. This transformation mirrors broader shifts in the creative industries, where AI-assisted automation displaces traditional creative workflows and restructures the role of human labour. As AI systems become primary agents of branding production, creative authority shifts from designers and art directors to algorithmic frameworks, creating a branding landscape that is increasingly shaped by machine-learning models rather than by individual artistic vision.

Beyond image generation, AI's influence extends to the automation of branding labour itself. Lau notes that AI reduces branding production costs by up to 80%, replacing traditional campaign shoots with synthetically generated content (Lau, 2023). A parallel trend in the media industries, observing that Generative AI is accelerating a transition toward asset-light digital workflows, where branding production becomes an exercise in algorithmic curation rather than a human-led creative process (Bender, 2024). While this shift enhances efficiency, it also diminishes the role of improvisation and human intuition in branding, favouring highly optimised, machine-generated branding aesthetics. Lau has expressed similar concerns, noting that AI-generated models often lack the distinctive imperfections that define human-based branding, contributing to a growing sense of hyper-uniform, computationally aestheticised brand imagery (Lau, 2024).

AI's integration into branding thus marks a profound transformation in the labour structures, aesthetics, and creative processes that underpin digital fashion branding. This shift necessitates a re-evaluation of what constitutes "creative work" (Bender, 2024; Erickson, 2024) in an AI-driven economy, as branding becomes a process of computational refinement rather than an exercise in traditional artistic authorship. While AI enhances branding efficiency, it also reconfigures the notion of artistic agency, shifting creative decision-making from human-led artistry to algorithmically mediated branding frameworks. Whether AI will remain a recursively modern tool for creative optimisation or evolve into an agent of radical aesthetic disruption remains an open question.

3.4 CONCLUSION

AI is redefining fashion branding, reshaping creative production, branding strategies, and labour structures. The shift from metaverse-driven digitalisation to AI-enhanced branding signals a movement away from static digital personas toward dynamic, data-driven marketing automation. Lau's DESIGNATOR AI Studio exemplifies this shift, demonstrating how AI-generated outputs function as iterative extensions of existing brand identities rather than radical departures from traditional aesthetics.

Grounded in recursive modernity, this study highlights how AI-driven branding operates within a continuous feedback loop, where branding aesthetics evolve through recombination rather than abrupt transformation. Generative AI does not disrupt brand identity but refines it, reinforcing the recursive evolution of fashion aesthetics through tools such as MidJourney and Stable Diffusion. While AI has lowered creative barriers, expanded scalability, and enhanced efficiency, it also presents challenges of aesthetic standardisation, shifting creative agency, and the potential erosion of human-led artistic spontaneity.

AI's impact on branding is neither wholly disruptive nor entirely subordinate to human creativity. Rather, it functions as an adaptive mechanism that refines, reinterprets, and extends existing aesthetic paradigms. As brand communication increasingly integrates generative AI, the industry enters an era of algorithmic co-creation, where human intuition and computational precision interact in a recursive process of aesthetic evolution.

BIBLIOGRAPHIES AND REFERENCE LISTS

Afifah, N. (2024) 'Influencer marketing strategies and the use of AI technology in enhancing brand awareness: A case study in the fashion industry', *Journal of Economic, Business and Accounting (COSTING)*, 7(2), pp. 3058–3066.

Balmain (2023) *Balmain Unveils AI-Powered Project for Unicorn Sneakers*. Available at: https://fashionunited.uk/news/business/balmain-unveils-ai-powered-project-for-unicorn-sneakers/2023120873033 (Accessed: 17 October 2024).

Barthes, R. (1967) *The Fashion System*. Berkeley: University of California Press.

Beck, U. (1992) *Risk Society: Towards a New Modernity*. London: SAGE Publications.

Bender, S. (2024) 'Generative-AI, the media industries, and the disappearance of human creative labour', *Media Practice and Education*, pp. 1–18.

Bharati, P. and Rathore, A. (2017) 'Sustainable fashion marketing: AI-powered solutions for effective promotions', *International Journal of New Media Studies*, 4(2), pp. 70–73.

Bratton, B. (2016) *The Stack: On Software and Sovereignty*. Cambridge, MA: MIT Press.

Bratton, B. (2021) *The Revenge of the Real: Politics for a Post-Pandemic World*. London: Verso.

Business of Fashion (2023) *Gen Z and the Rise of Virtual Fashion Influencers*. Available at: https://www.businessoffashion.com/reports/technology (Accessed: 20 June 2024).

Candeloro, D. (2020) 'Towards sustainable fashion: The role of artificial intelligence – H&M, Stella McCartney, Farfetch, Moosejaw: A multiple case study', *ZoneModa Journal*, 10(2), pp. 91–105.

Chen, X., Wang, Y. and Li, Z. (2020) 'The integration of AI and traditional craftsmanship in luxury fashion: A case study of Louis Vuitton', *Fashion Technology Journal*, 15(3), pp. 45–62.

Data Science Dojo (2023) *MidJourney 2023: AI-Generated Art Revolutionising Creativity*. Available at: https://datasciencedojo.com/blog/MidJourney-and-ai-art/ (Accessed: 20 April 2023).

Deng, L. (2022) 'Virtual models as the new brand ambassadors in fashion: A trend analysis', *Journal of Digital Branding*, 10(2), pp. 12–24.

Epic Games (2021) 'Balenciaga brings high fashion to fortnite', *Epic Games News*, 20 September. Available at: https://www.epicgames.com/site/en-US/news/balenciaga-brings-high-fashion-to-fortnite (Accessed: 25 September 2025).

Evangelista, P.N. (2020) *Artificial Intelligence in Fashion: How Consumers and the Fashion System are Being Impacted by AI-Powered Technologies*. Master's thesis, Politecnico di Milano. Available at: https://www.politesi.polimi.it/retrieve/38a8d87b-d3ae-4c34-b2ed-4259cced85a2/2020_10_Evangelista.pdf (Accessed: 18 September 2024).

FashionUnited (2022) *The Rise of Virtual Influencers: The Phenomena Taking Over Fashion Marketing*. Available at: https://fashionunited.uk/news/business/the-rise-of-virtual-influencers-the-phenomena-taking-over-fashion-marketing/2022122066897 (Accessed: 2 October 2024).

Fischer, M.M.J. (2018) 'Recursive modernity in science and technology studies', *East Asian Science, Technology, and Society*, 12(4), pp. 567–586. Available at: https://dspace.mit.edu/bitstream/handle/1721.1/123860/Theorizing%20STS%20from%20Asia.EAS12.4_Fischer_18752160-7219380_1pp_reduced.pdf?sequence=2 (Accessed: 14 October 2024).

Friedman, J. (1994) *Cultural Identity and Global Process*. Chicago: University of Chicago Press.

Giddens, A. (1991) *Modernity and Self-Identity: Self and Society in the Late Modern Age*. Cambridge: Polity Press.

Gielens, K. and Steenkamp, J. (2019) 'Branding in the age of digital disruption', *Journal of Marketing Research*, 56(5), pp. 643–659.

Guo, Z. et al. (2023) 'AI assisted fashion design: A review', *IEEE Access*, 11, pp. 88403–88415. https://doi.org/10.1109/ACCESS.2023.3306235

Huang, X., Kettley, S. and Lycouris, S. (2024) 'Developing shape change-based fashion prototyping strategies: Enhancing computational thinking in fashion practice and creativity', *Fashion Practice*, pp. 1–29.

Huyssen, A. (2003) *Present Pasts: Urban Palimpsests and the Politics of Memory*. Stanford, CA: Stanford University Press.

Influencer China Agency (2023) *The Fall of Virtual Influencer Ayayi in China*. Available at: https://influencerchinaagency.com/fall-virtual-influencer-ayayi-china (Accessed: 2 February 2025).

Ishizumi, M. (2022) 'Japan as a political community and the banner to raise', *International Business Review, Kanagawa University*, 3(1), pp. 45–67. Available at: https://kanagawa-u.repo.nii.ac.jp/?action=repository_uri&item_id=14600&file_id=18&file_no=1 (Accessed: 2 December 2024).

Jeon, Y. et al. (2021) 'FashionQ: An AI-driven creativity support tool for facilitating ideation in fashion design', in *Proceedings of the 2021 CHI Conference on Human Factors in Computing Systems*, pp. 1–18. New York: Association for Computing Machinery, Article 576. https://doi.org/10.1145/3411764.3445093.

Jing Daily (2023) *Web3: Mugler, Balmain, and Emerging Tech*. Available at: https://jingdaily.com/posts/web3-mugler-balmain-emerging-tech (Accessed: 29 December 2024).

Jing Daily (2024) *Gucci and Christie's Explore Generative Art, AI, and NFTs*. Available at: https://jingdaily.com/posts/gucci-christies-generative-art-ai-nfts (Accessed: 21 September 2024).

Jones, M. (2022) 'AI in Advertising: How Personalisation Drives Consumer Engagement', *Digital Marketing Review*, 9(2), pp. 23–37.

Kellner, D. (1988) 'Postmodernism as social theory: Some challenges and problems', *Theory, Culture & Society*, 5(2–3), pp. 239–269.

KDnuggets (2024) *Generative AI Playground: Text-to-Image Stable Diffusion with Stability AI, Stable Diffusion XL, and CompVis on the Latest Intel® GPU*. Available at: https://www.kdnuggets.com/2024/02/intel-generative-ai-playground-text-to-image-stable-diffusion (Accessed: 20 April 2024).

Kim, Y. (2022) 'Enhancing online shopping with AI', *E-Commerce Journal*, 18(3), pp. 45–67.

Kohli, C.S., Ebrahimi, M. and Granitz, N. (2023) 'Death bells? Branding in the age of smartphones and AI', *Journal of Business Strategy*, 44(6), pp. 399–405.

Lagerkvist, A. (2013) 'Retromodern Shanghai: Uncanny memories of media futures past', in *Media and Memory in New Shanghai: Western Performances of Futures Past*. London: Palgrave Macmillan, pp. 89–112. Available at: https://link.springer.com/chapter/10.1057/9781137014658_3 (Accessed: 9 November 2024).

Lau, P. (2023) *Semi-Structured Interview with Percy Lau*. Undertaken on 15th August, and 22nd August 2023 in Shanghai, by Zhe Wang and Jessica Yanna Xu.

Lau, P. (2024) *Semi-Structured Interview with Percy Lau*. Undertaken on 10th April, and 25th April 2024 in Shanghai, by Zhe Wang and Jessica Yanna Xu.

Lee, S. (2021) 'Predicting fashion trends with AI', *Journal of Fashion Technology*, 15(2), pp. 34–56.

Liang, Y., Lee, S.H. and Workman, J.E. (2020) 'Implementation of artificial intelligence in fashion: Are consumers ready?', *Clothing and Textiles Research Journal*, 38(1), pp. 3–18.

Luce, L. (2018) *Artificial Intelligence for Fashion: How AI is Revolutionizing the Fashion Industry*. New York: Apress.

Manovich, L. (2020) *AI Aesthetics*. Moscow: Strelka Press.

Martin, T. (2021) 'NFTs and fashion: A revolutionary combination in the digital age', *Digital Culture and Society*, 6(1), pp. 33–45.

Nadeem, R. (2023) 'Midjourney: Revolutionizing art with AI-generated creativity', *Data Science Dojo Blog*, 15 March. Available at: https://datasciencedojo.com/blog/midjourney-and-ai-art/ (Accessed: 25 November 2025).

NVIDIA (2023) *Generate Stunning Images with Stable Diffusion XL on the NVIDIA AI Inference Platform*. Available at: https://developer.nvidia.com/blog/generate-stunning-images-with-stable-diffusion-xl-on-the-nvidia-ai-inference-platform/ (Accessed: 25 April 2023).

Oosterom, E.B., Baytar, F. and Maher, M. (2023) 'Fashion marketing with virtual humans as influencers', in Sayem, A.S.M. (ed.), *Digital Fashion Innovations*. Boca Raton: CRC Press, pp. 163–178.

OpenAI (2024) *ChatGPT Can Now See, Hear, and Speak*. Available at: https://openai.com/index/chatgpt-can-now-see-hear-and-speak/ (Accessed: 20 April 2024).

Pillarisetty, R. and Mishra, P. (2024) 'Application and sustainability of AI and machine learning in the various stages of the value chain in online fashion retail: A conceptual review', *Sustainability in Marketing Practice*, pp. 131–140.

Pithers, E. (2023) *AI Virtual Models: How Fashion Brands are Embracing Synthetic Influencers*. Available at: https://www.theguardian.com/fashion/2023/apr/03/ai-virtual-models-fashion-brands (Accessed: 27 November 2024).

Rathore, B. (2017a) 'Exploring the intersection of fashion marketing in the metaverse: Leveraging artificial intelligence for consumer engagement and brand innovation', *International Journal of New Media Studies*, 4(2), pp. 51–60.

Rathore, B. (2017b) 'Sustainable fashion marketing: AI-powered solutions for effective promotions', *International Journal of New Media Studies*, 4(2), pp. 70–80.

Rathore, B. (2019) 'Chic strategies: Revolutionizing the industry through innovative fashion marketing', *International Journal of New Media Studies*, 6(2), pp. 23–33.

Refabric (2023) *Fashion AI and Inclusivity: Designing for a Diverse World*. Available at: https://blog.refabric.com/fashion-ai-inclusivity-diverse-world/ (Accessed: 3 December 2024).

Rietze, A. (2016) 'Project Muze: Fashion inspired by you, designed by code', *Google Europe Blog*, 2 September. Available at: https://blog.google/around-the-globe/google-europe/project-muze-fashion-inspired-by-you/ (Accessed: 25 September 2025).

Sayem, A.S.M. (2023) *Digital Fashion Innovations: Advances in Design, Simulation, and Industry*. Boca Raton: CRC Press.

Sayre, R. (2013) 'Preparing for preparedness: Security, disaster, and recursive modernity in contemporary Japan', *Yale University Working Papers in Anthropology*, 27(3), pp. 33–51.

Tomba, M. (2019) *Insurgent Universality: An Alternative Legacy of Modernity*. Oxford: Oxford University Press.

Ulrich, B. and Giddens, A. (1994) *Reflexive Modernization: Politics, Tradition and Aesthetics in the Modern Social Order*. Cambridge: Polity Press.

University of the Arts London (UAL) (2023) *MA Fashion Communication Curriculum*. Available at: https://www.arts.ac.uk/subjects/fashion (Accessed: 20 June 2024).

Varsha, P.S. et al. (2021) 'The impact of artificial intelligence on branding: A bibliometric analysis (1982–2019)', *Journal of Global Information Management*, 29(4), pp. 221–246.

VideoWeek (2023) *How AI is Revolutionising the Future of Digital Creativity*. Available at: https://videoweek.com/2023/05/09/how-technology-ai-is-revolutionising-the-future-of-digital-creativity/ (Accessed: 25 April 2023).

World Economic Forum (2023) *AI for Sustainable Fashion: Opportunities and Risks*. Available at: https://www.weforum.org/reports/ai-for-sustainable-fashion (Accessed: 20 June 2024).

Zhang, Y. and Kim, H. (2024) 'Ethical implications of synthetic media in branding', *Journal of Business Ethics*, 189(1), pp. 45–63.

Zhang, Y. and Liu, C. (2024) 'Unlocking the potential of artificial intelligence in fashion design and e-commerce applications: The case of MidJourney', *Journal of Theoretical and Applied Electronic Commerce Research*, 19(1), pp. 654–670.

Zou, X. and Wong, W. (2021) *Fashion After Fashion: A Report of AI in Fashion*. arXiv preprint arXiv:2105.03050.

4 Enhancing Fashion Design Education and Industry Practices with Artificial Intelligence

Brian Taylor

4.1 BACKGROUND

As artificial intelligence (AI) continues to infiltrate numerous parts of creative industries, the fashion sector is experiencing a significant transformation. Central to this shift are new AI-driven tools that streamline ideation, reduce costs, and widen access to sophisticated design capabilities. Among these emerging platforms is Midjourney, a cutting-edge AI text-to-image generator, which serves as a compelling example of how generative AI can augment fashion design processes for both educational and professional applications. Midjourney translates textual descriptions – referred to as prompts – into vivid visual representations, enabling the rapid generation of mood boards, fabric and color simulations, and even entire conceptual design collections. By bridging the gap between abstract ideas and tangible visuals, it promises to redefine traditional workflows in the fashion landscape.

Drawing on a case study informed by inspirations from popular culture and regional craft traditions, this chapter probes Midjourney's utility as both an educational resource and a professional design tool. It emphasizes how this technology can dismantle barriers often rooted in limited technical skillsets, scarce financial resources, and restrictive timelines. In academic settings, for instance, Midjourney provides a new avenue for students – particularly novices or those less confident in their hand-drawing skills and garment construction knowledge – to engage more fully with the creative process. Freed from the constraints of material costs, lengthy sample-making procedures, and advanced software proficiencies, learners can rapidly explore diverse aesthetics and refine their concepts. For industry professionals, the enhanced workflows facilitated by Midjourney can potentially compress lengthy product development cycles, reducing not only material waste and shipping costs but also the environmental footprint associated with traditional apparel sampling methods.

However, as the fashion world embraces these capabilities, it must also confront the limitations and ethical implications of AI-driven creativity. While AI presents efficiency and democratization benefits, it is not a universal solution. Issues such

 DOI: 10.1201/9781003594482-5

as authorship, originality, and bias loom large. Since models like Midjourney learn from vast datasets that may contain skewed cultural references or historical prejudices, there is a risk of perpetuating stereotypes or misrepresenting marginalized communities. Designers, educators, and policymakers must therefore approach the integration of AI tools with critical insight, ensuring that these technologies promote inclusivity, cultural sensitivity, and responsible use. This explores Midjourney's capabilities and constraints, suggesting pathways for integrating AI tools more thoughtfully into a rapidly evolving digital fashion landscape, encouraging greater inclusivity, deeper creativity, and more sustainable practices.

4.2 INTRODUCTION

The intersection of fashion design and AI has accelerated over the last decade, profoundly affecting creativity, production, and pedagogy. The incorporation of AI into fashion once primarily concerned logistical or predictive aspects – such as trend forecasting, inventory management, and automated production. More recently, a new generation of AI models has begun to influence the heart of fashion's creative process. Generative AI platforms, including text-to-image models like Midjourney, have the potential to transform early-stage ideation, evolving from mere efficiency tools into authentic co-creators of the design narrative.

Midjourney, launched in 2022 and refined through subsequent versions (notably Versions 5 and 6 in 2023 and 2024), epitomizes this shift. Traditional CAD systems require user input through specialized interfaces and often assume a baseline level of technical fluency. In contrast, Midjourney operates primarily through language. By inputting carefully crafted textual prompts, a designer can obtain lifelike imagery that encapsulates desired textures, silhouettes, and aesthetics within seconds. This mode of interaction is more intuitive for many creative practitioners, breaking down barriers to entry and encouraging experimentation. Midjourney, in this sense, can be seen as a creative "co-pilot," assisting rather than replacing the human designer and fostering a more playful, iterative design environment.

Yet, the emergence of such tools prompts a range of questions. Is creativity compromised if algorithms generate the initial visualizations? How do we maintain the integrity of fashion education when foundational skills like drawing or patternmaking might seem less crucial at the conceptual stage? How can we ensure that models trained on large image datasets do not inadvertently reproduce harmful stereotypes related to race, gender, body size, or cultural heritage? These questions reflect broader societal debates on AI ethics, authenticity, and cultural responsibility.

This chapter addresses these concerns through a structured examination of Midjourney's integration into fashion design. It presents a case study that details the process of using Midjourney to generate mood boards, fabric simulations, and a thematically cohesive fashion collection inspired by both popular culture and regional craftsmanship. By analyzing the outcomes, benefits, and limitations, this chapter seeks to inform educators, students, and industry stakeholders about best practices for incorporating AI into their work. Ultimately, the aim is to envision a future in which AI serves as a partner in creativity rather than a hindrance, enriching the educational landscape, expanding the industry's capabilities, and fostering a more thoughtful, ethical approach to design.

4.3 LITERATURE REVIEW AND CONTEXTUAL BACKGROUND

The integration of AI into creative sectors – from literature and film to product design and architecture – has been steadily increasing over the past decade. However, fashion, with its high dependence on visual aesthetics, tactile understanding, and cultural narratives, presents a unique arena for exploring AI's transformative potential.

Initial forays into AI for fashion design leaned heavily on data-driven insights. For example, AI has been used to predict upcoming trends by analyzing social media feeds, runway images, and retail metrics. While these applications enhanced efficiency and market responsiveness, they did not significantly impact the essence of the creative process itself. More recently, research has highlighted the potential of generative AI tools to influence ideation, a fundamental phase where designers define mood, concept, and narrative for upcoming collections (Zhang & Liu, 2023; Yin et al., 2023).

One of the key shifts in the literature is toward democratized creativity. Tools like Midjourney can lower technical barriers, allowing individuals with limited traditional training to engage meaningfully in design. Vartiainen (2021) argues that integrating AI into craft and design education prepares learners for a technologically advanced future, encouraging them to develop a hybrid skillset that combines creative thinking, prompt engineering, and critical technological literacy.

At the same time, the literature underscores the importance of addressing biases in AI outputs. Zhou et al. (2024) and Hugging Face (2023) have documented how AI models often reflect the cultural and social biases embedded in their training data. This is especially relevant in fashion, an industry that plays a significant role in representing – and shaping – public conceptions of beauty, identity, and inclusivity. If AI-generated imagery tends to exclude certain body types, ethnicities, or cultural aesthetics, these tools risk deepening existing inequalities.

Another emerging theme in the literature is sustainability. Fashion's well-documented environmental footprint, from excessive textile waste to carbon emissions in shipping samples worldwide, has prompted many designers and brands to seek greener alternatives. By facilitating rapid digital prototyping, Midjourney and similar AI tools can reduce the need for multiple physical samples, thereby cutting costs, waste, and emissions. Cheng (2021) and other scholars highlight that such tools could help transition the industry toward more environmentally responsible practices, aligning technological advancement with ethical stewardship.

Overall, the literature supports the idea that Midjourney's emergence sits at the intersection of several key trends: the democratization of creativity, the need for ethical and inclusive design, and the pursuit of more sustainable production models. By acknowledging both the potential and pitfalls of AI in fashion, the literature sets the stage for a layered, critical examination of how tools like Midjourney can be integrated into design education and professional workflows.

4.4 METHODOLOGY

This research employs an exploratory case study methodology to understand Midjourney's capacity to produce a cohesive, thematically driven fashion collection. Unlike broad surveys or large-scale quantitative studies, this approach focuses on one detailed scenario to capture the complexities and nuances associated with the tool's use. The case study's theme draws inspiration from two sources: Madonna's "Take a Bow" (1994) music video – rich with melancholic romance and Old-World elegance – and the artisanal, homespun craft traditions of the Southeastern United States. By blending high-profile cultural references with local craftsmanship, the project tests Midjourney's ability to handle multiple, layered aesthetic inputs that demand both conceptual subtlety and visual coherence.

4.4.1 Research Phases

1. *Conceptualization and Theme Selection*: The research began by defining a conceptual direction that merged vintage romanticism and artisanal handwork. The aim was to create a series of men's athleisure garments fused with feminine touches and delicate hand-sewn details. As seen in Figure 4.1, this juxtaposition required Midjourney to translate not just stylistic elements – such as lavish embroidery or earthy color palettes – but also conceptual notes about mood, narrative, and thematic cohesion.
2. *Prompt Engineering for Mood and Fabric Boards*: The first deliverables were mood boards and fabric simulations. Detailed textual prompts described the desired textures (e.g., homespun craftsmanship or handwoven wool), color schemes (soft browns, muted golds, and warm neutrals), and environmental atmospheres (garden of Eden, dilapidated interiors, and rocky sculpture background). Iteration played a key role, as initial prompts often required refinement to achieve the desired aesthetic. Adjustments included adding specific reference points (e.g., "Spanish matadors" or "homespun sewing techniques") and varying descriptors until the outputs aligned closely with the intended narrative.
3. *Iterative Design Generation*: Building on the mood and fabric boards, the next phase involved generating apparel prototypes. Midjourney responded to prompts detailing silhouettes ("a men's embroidered tank top") or material combinations ("merino wool sleeveless sweater with silk embroidered flowers"), as seen in Figure 4.2. Each iteration revealed the importance of prompt precision. Vague language produced unpredictable results, whereas clear and context-rich prompts yielded images that closely mirrored the conceptual vision.
4. *Visual Merchandising and Brand Identity Integration*: Beyond standalone garments, the research tested Midjourney's ability to deliver consistent visual narratives across branding touchpoints. Prompts requested runway

FIGURE 4.1 Collection of mood boards and fabrication boards.

(© MidJourney Version 5 and artist Brian Taylor)

scenes (Figure 4.3), editorial photoshoot, store layout, visual merchandising, and packaging concepts (Figure 4.4), and even accessories like footwear or headwear (Figure 4.5). The objective was to see if Midjourney could maintain thematic coherence across these diverse elements, effectively simulating a brand identity.

5. *Evaluation and Analysis*: The final step was a qualitative evaluation of Midjourney's outputs, focusing on visual coherence, fidelity to the theme, technical detail accuracy, and the ability to inform further design refinements. Each iteration was documented, including the exact textual prompts, parameter settings, and resulting images. Multiple versions of Midjourney

FIGURE 4.2 Merino wool sleeveless sweater with silk embroidered flowers.

(© MidJourney Version 5 and artist Brian Taylor)

were tested to understand the impact of updated models, which often included improved responsiveness to nuanced prompts. This iterative documentation allowed the researcher to develop best practices for prompt engineering and to identify patterns in Midjourney's strengths and limitations.

4.5 RESULTS

The research demonstrated that Midjourney can indeed serve as a powerful bridge between conceptual inspiration and visual realization. It accelerated ideation, enriched brand narratives, and offered a new dimension of flexibility to both students and professionals. Yet, it also highlighted areas where careful navigation, critical thinking, and skill development in prompt engineering are essential.

The research discovered that using more precise language was essential not only for capturing garment silhouettes and accurately depicting models, but also for articulating specific design details and setting an immersive environment. Multiple rounds of AI outputs demonstrated that highly detailed descriptors – such as accurate fabric names, explicit apparel product terms, and nuanced design elements, significantly enhanced Midjourney's ability to render the intended vision. Fine-tuning the tool's parameters, including aspect ratio and style weight, allowed for refined control

FIGURE 4.3 Runway scenes.

(© MidJourney Version 5 and artist Brian Taylor)

FIGURE 4.4 Editorial photoshoot, store layout, visual merchandising, and packaging concepts.

(© MidJourney Version 5 and artist Brian Taylor)

FIGURE 4.5 Accessories.

(© MidJourney Version 5 and artist Brian Taylor)

over the AI's interpretive style, while explicit reference images and careful scene descriptors further aligned outputs with the designer's concept. Collectively, these strategies highlighted the importance of detailed, context-rich prompts to translate abstract ideas into compelling, accurate visualizations.

4.5.1 Successful Prompt Examples

- Men's fashion show photography, wide shot, outdoor whimsical floral runway garden, garden of Eden theme, intricate architecture, botanical garden, male model walking on runway, wearing an unbuttoned hoodie embellished with floral homespun craftsmanship flower applique and crochet, in the style of orange, white, tan, wearing white shorts, tall fuzzy pink socks and white Converse shoes, hyper realistic, intricate sewing techniques, vogue runway shot, entire body is in frame, male model looking at camera, shot by David LaChapelle, 35mm lens, 8k, epic natural lighting, sharp focus –ar 16:9 –stylize 1000 –iw 0.5 –v 5
- Male model wearing large fuzzy bucket hat on the runway, embellished with homespun craftsmanship, crochet yarn, orange, white, tan, hyper realistic, intricate sewing techniques, detail, close up –ar 341:512 –stylize 1000 –iw 0.5 –v 5

- Fashion product photography, standing male model, wearing embroidered tank top::2 adorned with homespun craftsmanship, hyper realistic, look book shot, white backdrop, entire body is in frame, 35mm lens, 8k, epic volumetric lighting, sharp focus::1 –ar 1:1 –stylize 1000 –iw 0.5 –v 5
- Fashion product outdoor photography, standing male model with messy hair, wearing an embroidered anorak, pullover jacket with a high collar embellished with floral homespun craftsmanship and numerous 3D pockets, orange, tan, white shorts, tall socks and sneakers, hyper realistic, lookbook shot, rock sculpture background, rainy afternoon, entire body is in frame, lens, 8k, epic natural lighting, sharp focus –stylize 1000 –iw 0.5 –v 5

4.5.2 Enhanced Ideation and Visualization

One of the most immediate benefits observed was the speed and flexibility of the ideation process. Traditional workflows often involve assembling physical swatches, creating hand-drawn sketches, or navigating complex CAD systems. Midjourney allowed designers to iterate more freely, adjusting color palettes, fabrics, or silhouettes almost on the fly. In educational settings, this encourages students – especially those still developing manual skills – to experiment broadly. Freed from the pressure of producing perfect hand renderings, they can invest more creative energy in exploring new concepts and aesthetics.

4.5.3 Beyond the Garment: Branding and Storytelling

The ability to generate cohesive narratives extended beyond garment design. Midjourney successfully produced runway scenes that captured the intended mood, editorial spreads that showcased the garments in context, and even packaging and store concepts. This holistic approach helped illustrate that fashion design is about constructing an entire world or story, not just an isolated product. Students exposed to such exercises learn early on how brand identity, merchandising strategy, and marketing imagery intersect, making the design process more holistic and strategically informed.

4.5.4 Challenges in Accuracy and Technical Detail

While impressive, Midjourney's outputs were not consistently flawless. Certain intricate details – like complex embroidery patterns, unconventional garment drapes, or precise scaling – proved challenging. This highlighted a new skill area for designers: prompt engineering. Rather than passively accepting imperfect results, designers learned to refine their language, specify parameters, or include reference imagery as input. Out of 2,525 generated images, 74 generated images were used in the final presentation. Over time, students and professionals can become more adept at persuading higher fidelity outputs from AI tools, blending their artistic vision with a new form of linguistic and conceptual precision.

4.5.5 Navigating Bias and Ethical Considerations

A critical discovery was the potential for bias and stereotyping in AI-generated outputs. If the training data skewed toward particular beauty standards, body types, or cultural aesthetics, Midjourney's outputs could unintentionally reinforce these biases. In some instances, attempts to portray diverse body shapes resulted in images that subtly distorted the figure or normalized only a narrow range of proportions. During one session with Midjourney, 255 images were generated. Only 30 percent of those images contained a person of color as the subject. All subjects were male, specified by the prompt.

Addressing this issue requires ongoing vigilance. Students must be taught not only how to produce visually appealing results but also how to critically evaluate those results. Educators can integrate assignments that require students to identify and correct biased outputs, fostering an ethical awareness that is increasingly vital as AI gains ground in creative industries. The ethical dimension also encompasses questions of authorship. When AI assists in generating designs, who owns the final product, and how is intellectual property to be assigned or credited? Such questions are more than theoretical; they influence how designers engage with and trust the technologies at their disposal.

4.5.6 Environmental and Efficiency Benefits

From an industry standpoint, one of the most promising implications involves sustainability and cost efficiency. The traditional approach of producing multiple physical samples is both time-consuming and resource intensive. By producing accurate digital prototypes early in the process, teams can reduce the number of physical samples needed, thereby cutting down on material waste, water consumption, and transportation-related emissions. As fashion brands face increasing pressure to adopt sustainable practices, integrating AI into the design pipeline can yield tangible ecological benefits, without necessarily compromising on creativity or quality.

4.6 DISCUSSION

The integration of Midjourney into fashion design contexts suggests a powerful synergy between human creativity and machine intelligence. Designers can leverage AI not only to expedite ideation but also to explore broader, more inclusive aesthetic vocabularies. However, the transformative potential of this tool must be approached with care and a willingness to adapt educational practices, industry standards, and ethical frameworks.

4.6.1 Implications for Fashion Education

For fashion educators, Midjourney and similar tools present an opportunity to revitalize curricula. Traditional design programs often prioritize manual rendering, patternmaking, and sewing as initial competencies. While these skills remain crucial, introducing AI-based generative tools can complement them, broadening students' horizons at the conceptual stage. Students who struggle with technical drawing might find new confidence in expressing their ideas through well-crafted prompts. Those who excel in conceptual thinking over technical execution can gain more immediate feedback on their visions, encouraging them to push boundaries and iterate rapidly.

Educators must guard against overreliance. If students never learn foundational skills – understanding fabric behavior, garment construction, or pattern logic – they may become overly dependent on AI outputs. AI can simulate, but it cannot fully replace the tactile and experiential learning that comes from handling fabrics, fitting garments, or experimenting in a studio environment. A balanced approach, blending digital and traditional methods, can yield more holistic designers who are both creatively and technically adept.

4.6.2 Industry Adoption and Sustainable Practices

In the professional sphere, Midjourney can serve as an accelerator of innovation. Brands that quickly visualize concepts can respond more nimbly to market shifts, test ideas internally before public release, and refine their narratives. The reduction in sample production can also mean significant cost savings and a reduced environmental footprint. This aligns well with rising consumer expectations that fashion brands operate ethically and sustainably.

However, overreliance on AI at the concept stage might risk homogenizing aesthetics if designers fail to push beyond the AI's comfort zone. Much as template-based solutions in graphic design tools can lead to visual uniformity, AI prompts could drive designers toward certain recurring patterns or silhouettes if not actively challenged. The human element – critical thinking, personal vision, cultural insight – remains the bedrock of meaningful innovation.

4.6.3 Ethical Frameworks and Guidelines

The ethical aspect remains a priority. The fashion industry, long criticized for perpetuating narrow beauty standards, must ensure that AI tools do not exacerbate these problems. Designing with Midjourney should go hand in hand with developing guidelines for identifying and correcting biases in the outputs. Institutions might establish review committees or advisory boards to monitor AI-generated materials for inclusivity, requiring that final collections represent a diverse array of body types, cultural influences, and identities.

Educational programs could incorporate modules dedicated to AI ethics, encouraging students to consider the broader social implications of the imagery they produce. Such modules might cover the origin of training datasets, the impact of visual representation on marginalized communities, and strategies to diversify prompts. Over time, these efforts can help cultivate a generation of designers who see ethics not as an afterthought, but as an integral part of the creative process.

4.7 THE FUTURE OF AI IN FASHION

Midjourney is by no means the final stage of AI-driven creativity in fashion. We can anticipate more advanced models that allow real-time interaction, integrate with augmented or virtual reality systems, and simulate textures, draping, and fit with even greater realism. Designers may soon guide AI models through voice commands or haptic feedback, making the creative process more immersive and intuitive.

Future research might explore linking AI-driven concept generation directly to on-demand manufacturing systems. A designer could iterate with AI until a final concept emerges, and then send those specifications directly to automated, local production facilities. This scenario could further streamline supply chains, reduce waste, and personalize fashion for individual consumers. Additionally, specialized AI tools might emerge to serve niche markets, such as adaptive or assistive apparel, expanding the reach of fashion's inclusivity beyond aesthetics and into functionality.

Yet, these advancements will also intensify the ethical, cultural, and educational debates. As AI grows more powerful, the risk of homogenization or cultural appropriation may increase if designers rely too heavily on models trained on Western-centric datasets. The responsibility to continually expand and diversify training data, update ethical guidelines, and ensure cultural sensitivity will remain vital. Designers, educators, and policymakers will need to collaborate closely to ensure that technological innovation does not come at the expense of authenticity and equity.

4.8 CONCLUSION

This chapter has demonstrated how Midjourney, an AI text-to-image generator, can serve as both a catalyst and structural support in the fashion design process. By enabling rapid ideation, expanding educational inclusivity, and contributing to more sustainable workflows, Midjourney exemplifies the potential of AI to enhance creativity rather than stifle it. Its benefits – accelerated concept generation, holistic brand storytelling, and reduced material waste – mark a significant step forward in integrating technology into the fashion world.

However, these opportunities also surface critical challenges. Issues of bias, authenticity, and ethical responsibility demand attention. Without thoughtful oversight, AI tools could inadvertently reinforce harmful stereotypes or overshadow the importance of material knowledge and craftsmanship. The role of educators, industry leaders, and policymakers is crucial in shaping best practices, setting standards, and cultivating an environment in which AI complements traditional skillsets rather than replaces them.

As the fashion industry continues to navigate the digital era, tools like Midjourney will likely become integral to the design process. The question is not whether AI will influence the future of fashion, but how it will be integrated in ways that honor the richness of cultural traditions, uphold ethical standards, and empower a new generation of designers to think creatively and responsibly. Mastering these tools – and understanding their limitations – can foster a more inclusive, sustainable, and imaginative future for fashion design, where human ingenuity and machine intelligence coexist in harmony rather than tension.

REFERENCES

Cheng, S.-H. 2021. "Impact of Generative Artificial Intelligence on Footwear Design Concept and Ideation." *Engineering Proceedings* 55 (1).

Hugging Face. 2023, March 15. "Ethics and Society Newsletter #4: Bias in Text-to-Image Models." *Hugging Face*. https://huggingface.co/blog/ethics-soc-4.

Vartiainen, H. 2021. "Using Artificial Intelligence in Craft Education: Crafting with Text-to-Image Generative Models." *Digital Creativity* 34 (1).

Yin, H., Zhang, Z., and Liu, Y. 2023. "The Exploration of Integrating the Midjourney Artificial Intelligence Generated Content Tool into Design Systems to Direct Designers towards Future-Oriented Innovation." *Systems* 11: 566.

Zhang, Y., and Liu, C. 2023. "Unleashing the Potential: Exploring the Impact of Artificial Intelligence on Fashion Design Using Midjourney." *ITAA Proceedings* No. 80.

Zhou, M., Prabhu, V. U., Kahembwe, E., Birhane, A., and Han, S. 2024. "Bias in Generative AI." *arXiv*. https://doi.org/10.48550/arXiv.2403.02726.

5 Artificial Intelligence and Ageism in Fashion

Morolake Dairo and Sharon Nunoo

5.1 INTRODUCTION: BACKGROUND

The global fashion industry has been heavily criticized for prioritizing certain fashion groups from various social categories, especially race, gender, geography, body image and social class (Gabriel 2023, 106–116). In recent times, there has been a clamor for more inclusivity in the fashion industry, with global fashion companies establishing EDI (Equity, Diversity and Inclusivity) departments and initiatives to drive change, with a principal focus on gender, race and body positivity. One of such is the annual size inclusivity report published by *Vogue* (Maguire et al. 2024). However, one area that often gets overlooked is age. The global fashion industry has long prioritized youth, with a focus on perfection ideals that exclude wrinkles, gray hair and aging skin (Narasimhan 2017). Twigg (2013) notes that age has been used in society as a symbol of power and domination to enforce inequality. Dress and clothing for the older generation is usually designed in conservative aesthetics and is quite distinct from the more edgy and "fashionable" collections designed for the youth. A flip through fashion magazines and literature will also reveal a bias towards the more youthful body. These ideals and biases are also mirrored on the fashion runway, which mostly features younger models (Dale 2024; Hsieh 2017).

Even in the virtual world, a quick skim at the list of virtual influencers on social media platforms like Instagram reveals young and usually "flawless" characters such as 19-year-old Lil Miquela (Jhawar et al. 2023). Currently, none of them are above 50 years of age. In 2022, Nigerian-based filmmaker and graphic designer Malik Afegbua, using Midjourney, an AI platform, in combination with photoshop, created a virtual fashion series featuring black older models labelled "The Elder Series" (CNN 2023; Habte 2023). The images show elderly people of African descent dressed in a mix of African and modern-inspired pieces and proudly embodying their "aging" form with clear evidence of wrinkles, facial lines, visible veins on hands, age-induced posture and graying hair (See Figures 5.1–5.9). There have been criticisms on the use of artificially generated images (Oppenlaender et al. 2023), but the creation, similar to Afegbua's AI-generated series, could be a tool to addressing issues such as ageism in the global fashion industry.

This study explores the following research questions: (1) How has the Elderly Fashion Series by Malik Afegbua utilized AI in addressing representation and recognition of the older demographic in the fashion industry? (2) How can AI technologies be explored to solve inclusivity issues such as ageism in the global fashion industry?

DOI: 10.1201/9781003594482-6

5.2 AGEISM

Ageism refers to biases, stereotypes and discrimination arising from age and is not exclusive to the older generation (WHO, 2021). However, this study is limited to ageism amongst the older generation within the context of fashion. The definition of "old age" is quite complex as it is multifarious and can be defined across different geographies, eras and social categories for different reasons. In social and government contexts, retirement usually signifies old age, but the retirement age varies across countries and can be between 55 and 65 years of age (Loretto et al. 2000). In the clothing and fashion industry, age is approached differently with the "ultimate fashion age" bordering between late teenage years and early 20s for women, while for men there is still some sort of concession as older male models and figures are still termed as "distingusished" in their graying years (Twigg 2013, 26; Swift and Steeden 2020, 24). Meanwhile, the beauty sector classifies the appearance of any fine line on the skin as a symbol of aging with the marketing of cosmetics and face creams that can erase any aging lines for mostly women in their early 30s (Twigg 2013, 49). In 2023, the global botox sector was estimated at $4.3 billion and is embraced by both young and old but more common amongst women (PR Newswire 2024; Rock 2017; Singh et al. 2015).

Across a range of industries, age is a factor in the labor market (Swift and Steeden 2020, 17). For front-facing jobs such as acting, journalism and the arts, younger folks, especially females, are preferred to older and more experienced counterparts for visual aesthetics (Dolan 2013, 342–351). Hence the "retirement" age could be between 40s and 50s, except there is a need for an "older" character. For other sectors like politics and higher management, age is usually embraced as it is linked to the "older the wiser" belief of older leaders and CEOs who have gained relevant experience over the years.

Yet, on a biological level, for women menopause is on the horizon usually around their 40s, coupled with bodily changes. For men, physical and mental signs of aging may set in later in time through graying of hair, posture, reduction in bone density, amongst others (WHO 2021). And in the health sector, there are reports that older patients face discrimination in health care and are usually excluded from research studies (WHO 2021).

Twigg (2013) highlights that the history of social dress codes show that "age-coded dress" has always been a part of dress history, with images from the 1920s showing older women retaining longer skirt lengths as compared to younger women, and embracing darker colors. For the older male folk, longer robes and darker colors were also used to differentiate their apparel from that of younger men. The media has also contributed to ageism as evident in the representation of older people in print, visual and digital advertisements. As research shows (Loos and Ivan 2018, 163–176), younger adults are usually showcased as more trendy and healthy as compared to older counterparts, contributing to the stereotype of "visual ageism".

Research has shown that age-related prejudices and stereotypes can result in negative health outcomes including increased risk of dementia and shorter life span for the elderly (Swift and Steeden 2020, 18). An increased and positive representation of the older demographic in both offline and digital platforms and spaces can help correct these prejudices.

5.3 INCLUSIVITY IN THE FASHION INDUSTRY: AGEISM

Over the years, the approach of the global fashion industry to inclusivity and diversity has faced increased scrutiny (Farra 2021). But there has been a major shift in terms of brands paying more attention to race, gender, body positivity and mental health in their creation and marketing efforts (Apeagyei 2007; Vogure Business 2024). Whilst in some respect these addresses aspects related to inclusivity and diversity, focus on the older demographic is largely missing. Brands seem more focused on Gen Z consumers, with luxury brands expanding their brand ethos and traditional marketing strategies to cater more to this demographic (Hobbs 2021). The older market holds a strong and stable potential for the global fashion industry, as Kennedy (2024) notes that they have real spending power and, despite stereotypes on their media consumption habits, can be reached through digital platforms and more creative marketing. According to Mintel (2022) three in ten women above the age of 55 found it difficult purchasing clothing that suited their age. Rather than celebrate the process of aging, fashion and beauty brands seem focused on selling "youthfulness" to the older demographic by marketing anti-aging solutions and products or ignoring the older demographics.

5.4 ARTIFICIAL INTELLIGENCE AND THE REAL WORLD: INCLUSIVITY AND DIVERSITY

The media contributed to the debate about the inclusivity and diversity gap in the global fashion industry (Lewis et al. 2011), with research showing that majority of the representations of older people in media usually revolves around negative stereotypes relating to age, that border around less attractive, inactive, poorly dressed, unskilled and are mostly riddled with stereotypes of older women as "nurturers" and older men as "more powerful". Yet, just like the youthful population, there is a wide variety of personas and market segments common to the older population that goes beyond "old retiree" or "caring grandparent" (Swift and Steeden 2020; Kennedy 2024).

Research from Shin and Lee (2023) show there is already disparity in the virtual universe, mimicking what is obtainable in the physical world of fashion. Majority of the virtual influencers currently developed seem to be aged between 18–26 years, depicting usually slender, youthful bodies free of blemish and aging factors. Sylvia, a virtual influencer, seems to be the only older virtual model on Instagram who was created to show the different stages of human life, advancing from adolescence to old age and eventually death (Shin and Lee 2023). Visual elements such as wrinkles, graying of hair, color of skin were used to depict the aging process before her final "death" in 2020 (Shin and Lee 2023). In terms of body positivity, most of the AI-generated models and influencers, who are mostly women, are created with the "ideal" body representative of curves and sensual physiques. A few of them such as Karolin XS (@karolinxs) and Ono (@Ono.gram) on Instagram embrace a more gender-fluid persona. Virtual models and influencers of color accounted for the smallest representation in the number of AI-generated models across the world, while Europe had the highest number. Representation of individuals with special needs remains limited, with Kami (@itskamisworld) being one of the few examples. Created by the

Diigitals in partnership with the Down Syndrome International, Kami is a virtual influencer and model diagnosed with Down Syndrome (Keegan 2022). The question is no longer if AI-generated images and videos should be embraced but how they can be further explored for good in a world that is already evolving to incorporate these technologies. Currently, there is an imbalance in representation of age, gender, body and race in the digital landscape.

In 2022, Malik Afegbua, a Nigerian filmmaker and artist, created AI-generated images exploring older-looking models. According to the creator (Malik Afegbua, Interview, May 3, 2024), the AI-generated series offers an opportunity to challenge society's marginalized view of the older demographic. His images are distinct because he mostly uses models of color from the African continent dressed in a mix of modern and African-Nigerian inspired fashion pieces depicted in a fashion runway environment. This study will explore the imagery generated by Malik and explore how generative AI was deployed from an age-inclusive perspective.

5.5 METHODOLOGY

A content analysis and in-depth interview approach was adopted for this study. The content analysis approach was used in analyzing and reviewing the images in the first Elder series created by Malik Afegbua. By using this approach, insights can be gained into how creators and companies can apply AI in the positive representation of different demographics. An in-depth interview with the creator of the series was conducted to gain more insights into the motivations and application of AI generative images in driving age-inclusivity.

5.6 DISCUSSION OF FINDINGS: CONTENT ANALYSIS OF IMAGES

A visual content analysis (Bell 2004) was adopted for this study with a focus only on the images generated by AI. Comments and number of impressions from the Instagram post were not included in this analysis. The first generative AI series featuring the "Elders" on Instagram was posted on December 8, 2022, with the caption "Fashion show for the seniors #Ai" on Malik's official Instagram page (@slickcityceo). In total, there were six male models and four female models in the post, all generated from AI and edited using photoshop (Malik Afegbua, Interview, May 3, 2024). Following the visual analysis of the images, the following was observed:

Physical attributes of models: All the models featured in this series displayed older people of the African race with evidence of gray curly hair and curly beards typical of the hair morphology of sub-Saharan African (Khumalo 2005). There was also evidence of physical characteristics common to aging, such as wrinkles, face lines, evident collarbones (due to reduction in bone density), graying hair, sloppy shoulders (see Figures 5.1–5.10) (WHO 2021).

No evidence of makeup: There was no visible addition of makeup or filters to the images included in this series, as evident in the wrinkles and lines on the face and exposed features (neck, hands) of the models.

Clothing: Majority (seven) of the models wore African-inspired clothing, with only two males garbed in clothing that can be described as Western-inspired suits and accessories (see Figures 5.1 and 5.3) and one of the females (see Figure 5.5) dressed in a flower-pattern Western-style suit. The African elements included in the clothing featured African-inspired print, silhouettes (wrappers, kaftans, head ties) and accessories (earrings, beaded necklaces and bracelets) representative of African heritage (see Figures 5.2, 5.4, 5.5, 5.7, 5.8, 5.10).

Environment: All the models were located in a "fashion runway" environment with all the other elements blurred out, allowing sole focus on the model in the image.

FIGURE 5.1 Image of older male model in formal suit from the Elder Series created with the generative AI Midjourney Pro licensed version.

(Afegbua, 2022, with permission)

FIGURE 5.2 Image of older female model in Kaftan from the Elder Series created with the generative AI Midjourney Pro licensed version.

(Afegbua, 2022, with permission)

FIGURE 5.3 Image of older male model in formal suit from the Elder Series created with the generative AI Midjourney Pro licensed version.

(Afegbua, 2022, with permission)

FIGURE 5.4 Image of older male model in traditional attire from the Elder Series created with the generative AI Midjourney Pro licensed version.

(Afegbua, 2022, with permission)

FIGURE 5.5 Image of older female model in print jacket from the Elder Series created with the generative AI Midjourney Pro licensed version.

(Afegbua, 2022, with permission)

FIGURE 5.6 Image of older female model in formal suit from the Elder Series created with the generative AI Midjourney Pro licensed version.

(Afegbua, 2022, with permission)

FIGURE 5.7 Image of older male model in traditional top and bottom from the Elder Series created with the generative AI Midjourney Pro licensed version.

(Afegbua, 2022, with permission)

FIGURE 5.8 Image of older male model in traditional top and bottom from the Elder Series created with the generative AI Midjourney Pro licensed version.

(Afegbua, 2022, with permission)

FIGURE 5.9 Image of older male model in African-inspired suit from the Elder Series created with the generative AI Midjourney Pro licensed version.

(Afegbua, 2022, with permission)

FIGURE 5.10 Image of older female model in traditional attire from the Elder Series created with the generative AI Midjourney Pro licensed version.

(Afegbua, 2022, with permission)

5.7 DISCUSSION OF FINDINGS: IN-DEPTH INTERVIEW

In understanding the use of generative AI as a tool for inclusivity, an interview with Malik Afegbua was conducted in 2024 to explore the inspiration, implementation and impact of his use of generative AI in creating this imagery centered around the African elderlies and fashion. Malik Afegbua, who runs a production company in Lagos, Nigeria, and is a business graduate of the University of Surrey, describes himself as a "creative technologist and storyteller across different mediums of expression," He states,

> I do many other things in the industry. But in the film industry, I tell stories. In the tech industry, I use virtual reality to tell stories as well and that is just looking for the best means or the best way I could use to reach people. And when I mean people, I mean, like as much people like a diverse group of people as well. So, I look for technology I find tech to either make my process faster or to inspire me to think of new ways of how to use this.
>
> (Malik Afegbua, Interview, May 3, 2024)

This section explores some of the insights generated from the interview.

Lack of age-diverse content in fashion: While his exploration of the use of AI was purely experimental, he notes that the elder series was inspired by personal circumstances after the illness of an elderly family member. His search for inspirational imagery around elderly people proved futile. He states:

> I was researching on fashion, you know, for the elderly people all over. I couldn't find anything. Literally nothing. So, I said, you know, let's do something about it. Let's create something. That's where that started from.
>
> (Malik Afegbua, Interview, May 3, 2024)

Inclusivity and diversity: In the analysis of the images from the elder series, there is a focus on elderlies from the black African race, with a focus not just to create one super influencer identity but to embrace different relatable personas amongst the older demographics. Malik explains the notion behind this imagery, noting the gap in positive representation of older black people in Africa. In his interview, he stated,

> I create stuff that I kind of mixed, you know, cultures and all of that for different other reasons. But my focus is . . . in Africa, we don't have the right narrative. It's kind of controlled or misinterpreted or mis narrated or something. And our history has never been documented, and even the ones that have been documented are mostly about the suppression of, you know, our race. . . . You know, slavery and everything, all of that. The ones that we could find, a fragment of what used to be. So, because of that, that has really affected the mindset of a lot of people, you know, in terms of asking for permission to do everything that access just really in the mindset and I know the power of storytelling, storytelling could shift mindsets. That's what Hollywood has been doing since inception. Very intentional, you know, with the kind of story they tell that could create a trend or, you know, make people do certain things.
>
> (Malik Afegbua, Interview, May 3, 2024)

Through the use of fashionable dress and accessories, Malik's designs give his male and female models a fashionable twist adopting avantgarde designs from designers such as Yohji Yamamoto.

The role of storytelling: There are numerous criticisms for generative AI and the danger it poses to the global creative industries. Malik is focused on the positive influence of AI to change negative perceptions about the African continent. He notes, "My plan is to always just use story-telling to do that and the easiest form of storytelling for me is AI because the access is just out there and the capabilities as well is mind blowing." In terms of some of the challenges he has faced, he notes,

> At first I was getting really biased images, very stereotypical African impoverished images, and I was really wondering why. . . . So I had to train it with my datasets for it to actually understand that, OK, this is what we look like. This is how an African should look like on a stage, not what you know, the narrative is on the, you know, on the Internet and all of that. So that's started from there. And I just delved deeper into it and understood that in Africa, we are literally at the back of the line. And that's dangerous. So we need to be at the forefront to correct this false narratives.
>
> (Malik Afegbua, Interview, May 3, 2024)

And while there are general mixed reviews about generative AI images, Malik mentions that most of the responses have been positive. He mentions:

> Everybody, everybody, everyone has a different meaning, a different opinion, personalised, but the general message I get every time is I'm looking forward to getting old and even more accolades from the elderlies who are mostly above 50s represented in these images.
>
> (Malik Afegbua, Interview, May 3, 2024)

Phygital results: One of the major impacts that the Elder series has resulted into are real fashion shows dedicated to the Elderlies. The outfits on the real runway were all inspired by the AI-generated outfits worn by his virtual models from his previous digital runways (Lee 2023; Lutterodt 2023) with outfits created in partnership with real-life fashion designers, thereby creating a co-creation eco-system of fashion labels and the older demographcis. Malik notes:

> We had two fashion shows last year. We had one in Nigeria and we had one in in the Netherlands and they were both massive like . . . and the reception afterwards were like, "when is the next one? When is the next one?" So even the models, the elderly models, were so excited to see themselves on stage, you know, on a catwalk show and all of that.
>
> (Malik Afegbua, Interview, May 3, 2024)

His work has also led to the nurturing of a community of elderly people who occasionally contribute to his creative input and are sometimes part of the physical runway shows linked to the series.

5.8 CONCLUSION

The elder series offers a useful case on the rise of AI generative images and the numerous opportunities it poses for EDI conversations in the global fashion industry. It also shows the possibility of the co-existence and co-creation of virtual and physical realities that could be a useful tool in other industries as well. The role of storytelling is also a crucial factor in creating authentic digital experiences that can be impactful to societies.

It also offers brands within and outside the fashion sector the opportunity to rethink marketing strategies for older demographics. This can potentially change the narrative for some brands, as most of their loyal customers fall within this generation, and they can tailor marketing efforts towards meeting their needs. It can also be used as a tool for social marketing in addressing issues encountered by the older demographic across the world. Through storytelling, AI-generated imagery can be used in correcting such stereotypes with the use of more inclusive and diverse imagery that represents the diversity of this demographic.

This study contributes to the growing literature on AI and fashion innovation in relation to age-inclusivity studies and its connection to media and representation studies. It enriches the scholarship on African fashion, technology and representation in global fashion narratives. This study encourages further academic inquiry into alternative approaches for redefining beauty and industry norms, especially amongst the elderly generation.

REFERENCES

Apeagyei, Phoebe. "Significance of Body Image Among UK Female Fashion Consumers: The Cult of Size Zero, the Skinny Trend." *International Journal of Fashion Design, Technology and Education* 1(1). https://doi.org/10.1080/17543260701867697.

Bell, Philip. 2004. "Content Analysis of Visual Images." In *The Handbook of Visual Analysis*, edited by Theo Van Leeuwen and Carey Jewitt, 10–34. London: SAGE Publications Ltd. https://doi.org/10.4135/9780857020062.

CNN. 2023. "Nigerian AI Artist creates a Fashion Show for Elderly People." Accessed June 23, 2024 at https://edition.cnn.com/style/article/malik-afegbua-elderly-fashion-ai-art-spc-intl/index.html.

Dale, Alexandria. 2024. "Meet Qin Huilan, the 70-Year-Old Model Who Just Walked for Miu Miu at Paris Fashion Week." Accessed June 6, 2024 at https://www.cosmopolitan.com/uk/fashion/style/a60109734/qin-huilan-70-year-old-model-miu-miu/.

Dolan, Josephine. 2013. "Smoothing the Wrinkles: Hollywood, Old Age Femininity and the Pathological Gaze." In *The Routledge Companion to Media and Gender. Routledge Media and Cultural Studies Companions*. Routledge.

Farra, Emily. 2021. "This is What's Missing in Fashion's Inclusivity Movement." Accessed June 6, 2024 at https://www.vogue.com/article/whats-missing-in-inclusivity-movement-adaptive-fashion-disabled-community.

Gabriel, Michelle Blair. 2023. "The Economics of Exclusion: Why Inclusion Doesn't Fit Fashion's Business Model." In *Pioneering New Perspectives in the Fashion Industry: Disruption, Diversity and Sustainable Innovation*, edited by Elaine L. Ritch, Catherine Canning and Julie McColl. Emerald Publishing Limited.

Habte, Tarik. 2023. "The Nigerian AI Artist Reimagining a Stylish Old Age." Accessed March 11, 2024 at https://www.bbc.co.uk/news/world-africa–64260739.

Hobbs, Julia. 2021. "It's Official—Gucci is the Most Popular Luxury Brand for Gen Z on TikTok." Accessed June 6, 2024 at https://vogue.sg/gucci-tiktok/.

Hsieh, Text Vanessa. 2017. "The New Face of Reebok is 'China's Hottest Grandpa'." Accessed May 29, 2024 at https://www.dazeddigital.com/fashion/article/35326/1/the-new-face-of-reebok-is-china-s-hottest-grandpa.

Jhawar, Anand, Prashant Kumar and Sanjeev Varshney. 2023. "The Emergence of Virtual Influencers: A Shift in the Influencer Marketing Paradigm." *Young Consumers Insight and Ideas for Responsible Marketers* 24(5). http://dx.doi.org/10.1108/YC-05-2022-1529.

Keegan, Matthew. 2022. "How the World's First Virtual Influencer with Down Syndrome was Created." Accessed April 2, 2024, Accessed May 29, 2024 at https://www.campaignasia.com/article/how-the-worlds-first-virtual-influencer-with-down-syndrome-was-created/477935.

Kennedy, Joan. 2024. "Fashion's Golden Opportunity with Older Shoppers." Accessed February 4, 2024 at https://www.businessoffashion.com/articles/marketing-pr/fashions-golden-opportunity-with-older-shoppers/.

Khumalo, Nonhlanhla. 2005. "African Hair Morphology: Macrostructure to Ultrastructure." *International Journal of Dermatology* 44(1). https://doi.org/10.1111/j.1365-4632.2005.02805.x.

Lee, George. 2023. "Fashioning a New Narrative: Malik Afegbua Challenges Stereotypes." Accessed April 2, 2024 at https://thiscuriouslife.uknica.co.uk/fashioning-a-new-narrative/.

Lewis, Denise C., Katalin Medvedev and Desiree M. Seponski. 2011. "Awakening to the Desires of Older Women: Deconstructing Ageism within Fashion Magazines." *Journal of Aging Studies* 25: 101–109.

Loos, Eugene and Loredana Ivan. 2018. "Visual Ageism in the Media." In *Contemporary Perspectives on Ageism*, edited by Liat Ayalon and Clemens Tesch-Römer. Springer.

Loretto, Wendy, Colin Duncan and Phil White. 2000. "Ageism and Employment: Controversies, Ambiguities and Younger People's Perceptions." *Ageing and Society* 20: 279–302.

Lutterodt, Kai. 2023. "Africa Fashion Week Amsterdam (AFWA) Collaborates with New Silver Generation for a Fashion Show that Defies Age." Accessed April 2, 2024 at https://mysohotimes.co.uk/2023/08/16/africa-fashion-week-amsterdam-afwa-collaborates-with-new-silver-generation-for-a-fashion-show-that-defies-age-my-soho-times/.

Maguire, Lucy, Maliha Shoaib and Madeline Schulz. 2024. "The Vogue Business AW24 Size Inclusivity Report." Accessed March 11, 2024 at https://www.voguebusiness.com/story/fashion/the-vogue-business-aw24-size-inclusivity-report?uID=cf6b799c59e0217dffc4e75f889a6c8308f525d7922b8ce2ce3b4a9373612c64&utm_campaign=newsletter_weekly&utm_source=newsletter&utm_brand=vb&utm_mailing=VB_NEWS_Hilary_M.

Mintel Report. 2022. "Fashion-Inclusivity and Diversity – Uk 2022." Accessed February 4, 2024 at https://clients.mintel.com/content/report/fashion-inclusivity-and-diversity-uk-2022?fromSearch=%3Ffreetext%3DUK%2520Fashion%2520Inclusivity%2520and%2520Diversity%2520Report%25202022%26resultPosition%3D1.

Narasimhan, Divya. 2017. "Considering Your Age: A Study of How Women Above the Age of Sixty Express Themselves through Fashion." *Fashion – Scope: (Art & Design)* 15: 42–49.

Oppenlaender, Jonas, Johanna Silvennoinen, Ville Paananen and Aku Visuri. 2023. *Perceptions and Realities of Text-to-Image Generation*. Tampere: Mindtrek.

PR Newswire. 2024. "Botox Market Size to Hit USD 13.6 Billion by 2035 – Exclusive Report by Roots Analysis." Accessed March 11, 2024 at https://www.prnewswire.co.uk/news-releases/botox-market-size-to-hit-usd-13–6-billion-by-2035--exclusive-report-by-roots-analysis–302034691.html#:~:text=Global%20Botox%20Market%20Overview,period%20from%202023%20to%202035.

Rock, Lucy. 2017." Is America Developing a 'Crack-Like Addiction' to Botox Beauty?" Accessed March 11, 2024 at https://www.theguardian.com/lifeandstyle/2017/jan/08/botox-nation-america-addiction-crack-like-cosmetic-procedures.

Shin, Yeongyo and Selee Lee. 2023. "Issues of Virtual Fashion Influencers' Reproduced Bodies: A Qualitative Analysis Based on Body Discourse." *Fashion and Textiles* 10(30). https://doi.org/10.1186/s40691-023-00349-5.

Singh, Carter, Amarjit Dulku, Ahmed Haq, Tayyab Bhatti and Aqib Bhatti. 2015. "Why Do Females Use Botulinum Toxin Injections?" *J Cutan Aesthet Surg* 8(4): 236–238.

Swift, Hannah J. and Ben Steeden. 2020. *Exploring Representations of Old Age and Ageing*. Centre for Ageing Better.

Twigg, Julia. 2013. *Fashion and Age. Dress, the Body and Later Life*. Bloomsbury.

Vogure Business. 2024. "The Vogue Business Spring/Summer 2025 Size Inclusivity Report." Accessed February 4, 20224 at https://www.voguebusiness.com/story/fashion/the-vogue-business-spring-summer-2025-size-inclusivity-report.

WHO. 2021. *Global Report on Ageism*. Accessed March 11, 2024 at https://iris.who.int/server/api/core/bitstreams/71ad96a0-d29a-4457-9d54-52029c24c76c/content

Part B

Virtual Fashion Applications

6 Virtual Body and Satisfaction

Evrim Buyukaslan Oosterom, Fatma Baytar, and Fatma Kalaoglu

6.1 INTRODUCTION

Through social communication, individuals gain the ability to think carefully about their actions, consider the consequences of different choices, learn to see themselves as others do, and share common beliefs with others (Mead 1934). One such shared belief is the definition of ideal beauty (Orbach 1978). Cultural, artistic, and media-driven forces have influenced people's perceptions of ideal beauty, which evolved significantly over time (Calogero et al. 2007; Tan Ngo 2019). As noted by Dr. Valerie Steele, in the early twentieth century, the ideal beauty standard shifted from "Botticelli's opulent Venus to a slender, athletic Diana" (The Museum at FIT n.d.). Over the centuries, the media that propagate these beauty ideals have also transformed, transitioning from traditional paintings to print, broadcast, and social and digital media.

The concept of internalization explains how ideal beauty standards affect a person's body image. The thin female figure idealized today is a relatively new phenomenon. Although the slender female figure gained prominence in the nineteenth century, the thin-ideal beauty emerged only after the 1960s (Calogero et al. 2007). Over time, the concepts of internalization and the desire for thinness converged, leading to the development of thin-ideal internalization, which is the extent to which individuals adopt societal standards of thinness as their own aesthetic idea (Tiggemann and Lynch 2001). The widespread desire for a slimmer, idealized body reflects a societal fixation with appearance. Especially with the increase in social media use after 2008, most research has focused on the relationship between social media and body image (Vandenbosch 2022). Today, social media faces growing criticism for propagating unrealistic beauty standards that adversely impact self-esteem and body satisfaction (Möri et al. 2022; The Learning Network 2022). The rise of artificial intelligence (AI) has further complicated perceptions of beauty. Mobile applications now allow users to apply AI-powered filters that alter their facial and body features. Previous research has shown that the primary motivation behind the use of such filters is to conform to societal expectations of a slimmer and, thus, more attractive look (Maes and De Lenne 2022; Popescu 2022; Szambolics et al. 2023).

Additionally, with the increasing use of digital technologies, simulated or virtual bodies must be considered for their influence on body image and the evolving definition of ideal beauty (Deng and Jiang 2023). A survey conducted by Statista revealed that, in the U.S., 64 percent of Gen Zers were willing to create their digital

DOI: 10.1201/9781003594482-8

avatars for the metaverse (Statista 2022). With the advancements in AI and scanning technologies, people may soon interact with their digital avatars, which provide a more realistic and objective representation of their bodies than traditional mirrors (Arsenyan and Mirowska 2021). Therefore, considering the cultural and technological shifts, this chapter explores individuals' reactions to their three-dimensional (3D) body scan avatars and perceptions of their bodies. By doing so, the present study aims to shed light on the broader implications of integrating these technologies into consumer experiences.

6.2 BODY SATISFACTION

At its core, body satisfaction can be understood as the degree to which individuals feel content or discontent with their physical appearance. One of the earliest studies regarding this concept found a strong correlation between body satisfaction and self-esteem (Secord and Jourard 1953). Early studies often referred to body satisfaction as body cathexis, emphasizing its alignment with the cultural perception of ideal beauty. Numerous studies have demonstrated that pursuing societal beauty norms can have profound psychological consequences. Women experience heightened dissatisfaction when their bodies deviate from unrealistic standards (Benton and Karazsia 2014; Hawkins et al. 2004; McComb and Mills 2022). Another critical area of research centers on the relationship between body satisfaction and health-related behaviors. Studies have consistently shown that low levels of body satisfaction are predictors of unhealthy habits, such as smoking, avoiding physical activity, and developing eating disorders (Neumark-Sztainer et al. 2006; Cromley et al. 2012). Low body satisfaction often stems from internalized ideals that create an unattainable benchmark for many individuals.

Another important concept that should be considered in understanding body image is objectification. According to the objectification theory, people tend to see their bodies as objects to be observed by others if societal norms emphasize the importance of appearance. Internalizing other people's views leads to self-objectification (Fredrickson and Roberts 1997), which is correlated to body image (Muehlenkamp and Saris-Baglama 2002; Slater and Tiggemann 2002) and mediates the relationship between thin-ideal internalization and body satisfaction when people are exposed to thin-ideal appearances (Myers and Crowther 2007). Body satisfaction has also received attention in fashion research (Chattaraman and Rudd 2006; Grogan et al. 2013; Shim et al. 1991; Tiggemann and Lacey 2009). In 2018, the Fashion Institute of Technology (FIT) hosted an exhibition titled "The Body: Fashion and Physique," which communicated how societal expectations around clothing sizing contribute to body dissatisfaction. Curator Dr. Emma McClendon noted that rather than attributing the fit issues to systemic problems in the fashion industry, individuals often blame their bodies when clothing does not fit (The Museum at FIT 2018). Academic research supports this perspective, showing that inconsistent sizing standards lead to dissatisfaction and influence self-perception negatively (Apeagyei 2008; Grogan et al. 2020; LaBat and DeLong 1990).

Advancements in digital technologies have introduced new dimensions regarding body satisfaction. Virtual environments increasingly feature AI-generated virtual avatars as influencers, shifting the societal focus from real bodies to digitally constructed ones (Oosterom et al. 2023). Simultaneously, innovations such as virtual try-on technologies and fitness-focused mobile applications have allowed users to view highly detailed 3D representations of their bodies (Idrees et al. 2023). This experience differs fundamentally from the traditional act of looking in a mirror, offering a more detached and arguably objective perspective on one's appearance. As these technologies become more common, it is important to understand if they would foster greater acceptance of the body or amplify insecurities by highlighting perceived flaws. The following section delves into virtual bodies' psychological and emotional implications and examines how they impact body satisfaction in an increasingly digital world.

6.3 BODY VISUALIZATION TECHNOLOGIES AND THEIR RELATION TO BODY SATISFACTION

3D body scanning has introduced new aspects of self-awareness, allowing users to view highly realistic representations of their own bodies. Over the years, this technology has undergone remarkable advancements. Initially, 3D body scanners relied on white light or laser technology within large and stationary setups and were used primarily for research (Daanen and Ter Haar 2013). However, mobile applications have revolutionized accessibility, making 3D body scanning a tool available to the public (Idrees et al. 2023). Applications such as MagiScan (https://magiscan.app) can enable users to scan their bodies with a mobile phone equipped with a camera. The process involves capturing images from multiple angles while walking around the subject to be scanned. These images are then processed to generate a detailed, colored 3D body model, which can be downloaded in OBJ, STL, or FBX formats. Unlike a 2D mirror reflection, users can view the scans from a 360-degree perspective, despite minor distortions in areas such as the underarms and between the legs due to missing surface information.

The shift in how individuals view their bodies presents opportunities related to the technology's impact on body perception, psychology, health behaviors, and satisfaction with clothing. With the increasing adoption of 3D body scanning technology, researchers have investigated its psychological and behavioral effects on body satisfaction. However, findings in this area remain inconclusive. Although some studies suggested that viewing one's own 3D body could positively impact body satisfaction and promote acceptance and appreciation (Grogan et al. 2020; Neyret et al. 2020), other studies indicated that such exposure could intensify body dissatisfaction, particularly when individuals compare their 3D scans to societal beauty ideals (Grogan et al. 2015; Ridgway 2017; Park 2018). The discrepancy in findings emphasizes that body image has a complex nature, particularly in the context of emerging digital technologies. Self-esteem and self-compassion also play a significant role in understanding body image (Lafever et al. 2019; Park and Ogle 2021). Lafever et al. (2019) found that the self-esteem of college students, especially women, dropped after seeing their 3D body images and concluded a short-term negative impact. Park and Ogle (2021) investigated the effect of viewing 3D body scans on the self-esteem and self-compassion of young women and suggested to use 3D body scanning technology to enhance people's self-acceptance with body image.

Beyond its business-to-business (B2B) applications, the adoption of 3D garment simulation technology in consumer markets represents a significant transition. When consumers use these technologies, they are confronted with unaltered and realistic representations of their bodies, unlike the idealized, digitally modified appearances often presented through social media filters. This experience may provoke comparisons between their physical reality and societal beauty standards, potentially leading to decreased self-confidence and heightened discomfort (Möri et al. 2022). Given these implications, the current study sought to contribute to the body of research on the effects of viewing realistic virtual representations of one's body on body satisfaction. Therefore, our research questions (RQs) were as follows:

RQ1: Is there a difference between body satisfaction measured before and after women view their body models, and if so, which body areas contribute to this change?
RQ2*:* What is the relationship between BMI and changes in overall body satisfaction after individuals view their 3D body models?
RQ3*:* Is there a relationship between BMI and individuals' comfort levels after viewing their 3D body models?

6.4 METHODS

After receiving Cornell University Institutional Review Board approval (Protocol ID: 1712007670), women between 18 and 35 years old participated in the study in a laboratory setting. Following the definition of the U.S. Centers for Disease Control and Prevention (CDC), participants in the healthy-weight BMI category (i.e., 18.5–24.9 kg/m2) were recruited. Specific populations, such as plus-size individuals and older adults, may be more prone to experiencing decreased body satisfaction after viewing a 3D body scan. However, examining whether a similar effect occurs in young women with a healthy BMI is especially important, as this group often aligns more closely with societal beauty standards. Consequently, the present study focused on this demographic to better understand how 3D body scans might influence body satisfaction in this population.

Upon arrival at the lab, participants were scanned in their underwear with a Human Solutions Vitus XXL model 3D body scanner. Body satisfaction before (Real Body Satisfaction, RBS) and after (Virtual Body Satisfaction, VBS) seeing body scans was measured by using a seven-point Likert-type scale adopted from Shin and Baytar (2013). Likert-type scale was used to ensure statistical verifiability of the difference between body satisfaction before and after viewing the body scans. The body satisfaction scale items were anchored from 1, "very dissatisfied," to 7, "very satisfied." They measured the body satisfaction level for 22 items related to body parts (e.g., legs, waist, hips) and appearance (e.g., posture, body build). While the participants completed the survey after scanning and before viewing their 3D body scans, their scans were processed in Geomagic Wrap software. This involved smoothing, scaling, and saving the scans as.obj files. Participants were subsequently presented with their 3D body models on a computer screen (Figure 6.1), allowing them to interact with their virtual representations by zooming in, zooming out, and rotating the models. After this interaction, they completed the body satisfaction survey again. The absence of color in the body scans was a deliberate decision, as it ensured that participants focused solely on the shape and structure of their virtual body representations rather than being influenced by skin tone, clothing, or other visual details. The comfort level experienced by participants after viewing their own 3D body model was assessed using a seven-point scale adapted from Loker et al. (2004). The scale ranged from 1, "very uncomfortable," to 7, "very comfortable." The data was saved in a Microsoft Excel file and then analyzed and visualized using Excel and Python.

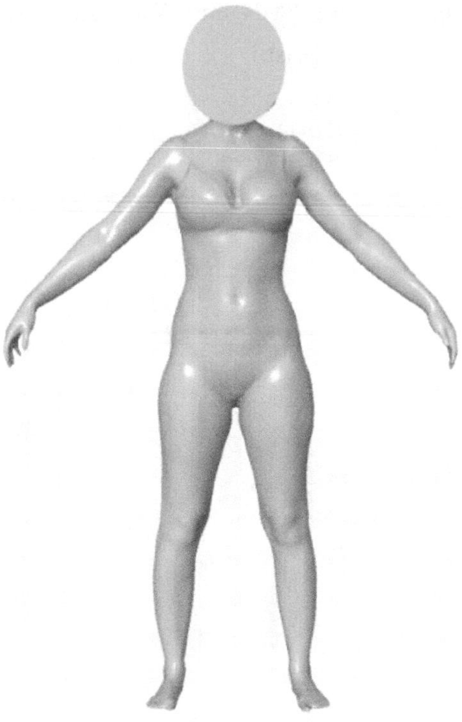

FIGURE 6.1 3D body scan of a study participant obtained by using Human Solutions Vitus XXL model 3D body scanner.

6.5 RESULTS

In total, 47 women participated in the study. The average age of the participants was 23.5 years, and the participants' average BMI was 22.04 kg/m^2. Descriptive statistics are reported in Table 6.1.

Figure 6.2 and Table 6.2 illustrate comparisons between the mean scores of RBS and VBS for each item on the body satisfaction scale. After viewing their 3D body scans, participants reported decreased body satisfaction for 15 items: appearance, waist, hips, body build, thighs, weight, abdomen, legs, buttocks, wrist, neck, height, back, profile, and torso. In contrast, increased body satisfaction was noted for seven items: shoulder width, feet, ankles, knees, arms, posture, and bust. The most significant changes in body satisfaction (VBS -RBS) were observed for body build (−0.77), appearance (−0.74), abdomen (−0.72), waist (-68), weight (−0.58), torso (−0.39), thighs (−0.36), and hips (−0.36), respectively. Conversely, the least amount of changes in body satisfaction were observed, in order, for height (+0.03), ankle (+0.05), and feet (+0.07), respectively. A paired t-test analysis was conducted

TABLE 6.1
Study Participants' Ages, BMIs, and Waist Girth Measurements

	Mean (SD)	Minimum	Maximum
Age	23.5 (4.7)	18	35
BMI [kg/m$^{2]}$	22.04 (2.0)	18.6	25.8
Waist Girth [cm]	71.0 (2.7)	62.2	78.5

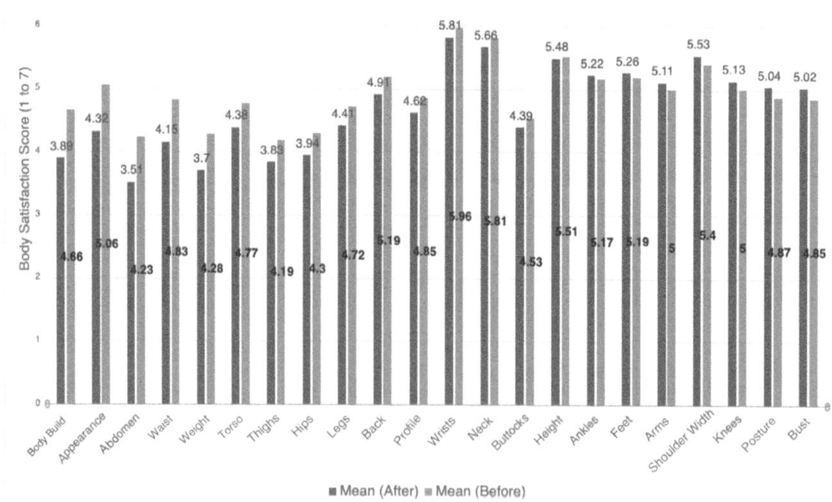

FIGURE 6.2 Comparisons between body satisfaction levels before (RBS) and after (VBS) seeing the 3D body model.

TABLE 6.2
Independent T-Test Analysis of Body Satisfaction Before (RBS) and After (VBS) Seeing 3D Body Scans

Item	M_{VBS}	SD_{VBS}	M_{RBS}	SD $_{RBS}$	t	p
Neck	5.66	1.17	5.81	1.16	0.71	0.48
Bust	5.02	1.69	4.85	1.57	−0.51	0.61
Torso	4.38	1.61	4.77	1.25	1.22	0.23
Shoulder Width	5.53	1.33	5.40	1.39	−0.45	0.65
Arms	5.11	1.45	5.00	1.29	−0.38	0.71
Wrist	5.81	1.15	5.96	1.03	0.57	0.57
Waist	4.15	1.58	4.83	1.34	2.40	0.02*
Abdomen	3.51	1.44	4.23	1.33	2.53	0.01*
Back	4.91	1.63	5.19	1.30	0.91	0.36
Hips	3.94	1.65	4.30	1.31	1.11	0.27
Buttocks	4.39	1.65	4.53	1.57	0.45	0.66
Thighs	3.83	1.68	4.19	1.58	1.20	0.23
Legs	4.41	1.75	4.72	1.60	0.80	0.43
Knees	5.13	1.54	5.00	1.37	−0.57	0.57
Ankle	5.22	1.60	5.17	1.55	−0.20	0.84
Feet	5.26	1.37	5.19	1.31	−0.23	0.82
Height	5.48	1.43	5.51	1.31	0.00	1.00
Weight	3.7	1.85	4.28	1.43	1.62	0.11
Profile	4.62	1.70	4.85	1.17	0.78	0.44
Posture	5.04	1.70	4.87	1.27	−0.55	0.58
Body Build	3.89	1.80	4.66	1.35	2.40	0.02*
Appearance	4.32	1.66	5.06	1.24	2.09	0.04*

Note. $^{*}p < 0.05$

to examine if these differences were statistically significant. Findings indicated a significant body satisfaction decrease for waist (t = 2,40, $p < 0.05$), abdomen (t = 2,53, $p < 0.05$), body build (t = 2,40, $p < 0.05$), and appearance (t = 2,09, $p < 0.05$) after seeing 3D body scans (Table 6.2).

In addition to analyzing body satisfaction for individual body parts, overall body satisfaction (i.e., the average of body satisfaction obtained for individual body parts) was also calculated. Even though the overall RBS was higher (M = 4.93, SD = 0.85) than the overall VBS (M = 4.70, SD = 1.05), the difference was not significant (t=1.13, $p = 0.26$). A correlation analysis was conducted to understand the relationship between participants' waist and abdomen circumference measurements and their corresponding body satisfaction levels in these areas. The correlation between waist girth and waist satisfaction change was positive but statistically insignificant

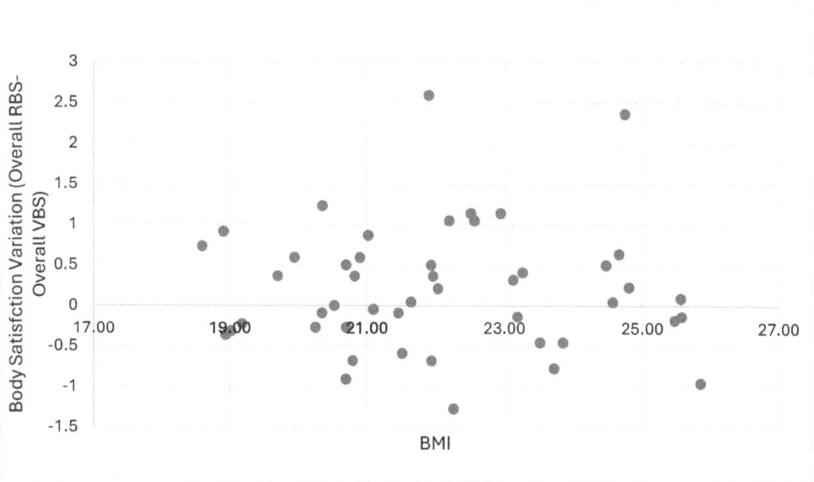

FIGURE 6.3 The correlation analysis between participants' BMI and their overall body satisfaction changes after seeing their 3D body model.

(r (45) = 0.21, p = 0.17). Similarly, the correlation between abdominal girth and abdomen satisfaction change was positive but not statistically significant (r (45) = 0.09, p = 0.57).

To address RQ2, Pearson correlation coefficients were calculated to examine the relationship between participants' BMI and the difference between RBS and VBS. As illustrated in Figure 6.3, the analysis revealed a positive but statistically insignificant correlation between BMI and overall body satisfaction change (r (45) = 0.04, p = 0.77). The findings regarding participants' comfort levels when viewing their own 3D body scans revealed that, on average, participants reported being comfortable with the experience (M = 5.65, SD = 1.46). Seventeen participants reported being very comfortable (score = 7), and 13 indicated they were comfortable (score = 6). Conversely, only one participant reported being uncomfortable (score = 2), and six participants expressed discomfort (score = 3) when viewing their 3D body scans.

To address RQ3, Pearson correlation coefficient analyses were conducted to examine the relationship between BMI and comfort levels in viewing one's own 3D body model. Although the analysis revealed a positive correlation, it was weak and statistically insignificant (r (45) = 0.01, p = 0.94). The average BMI of participants who reported being very comfortable (score = 7) or comfortable (score = 6) was calculated as M = 21.90 kg/m^2 (SD = 2.01). In comparison, participants who reported being uncomfortable (score = 2) or somewhat uncomfortable (score = 3) had an average BMI of 21.98 kg/m^2 (SD = 2.50).

6.6 DISCUSSION AND CONCLUSIONS

As virtual try-on technologies for online shopping become increasingly common (Idrees et al. 2023; WGSN Insider 2022), it is important to understand its effects on users' body perceptions and psychological well-being. Women increasingly continue to internalize societal pressure to meet the "thin-ideal" standard of beauty, which reinforces self-objectification. Their reactions to hyper-realistic 3D simulations will be a significant concern as 3D modeling and 3D body scanning technologies advance in the coming decades. The present study aimed to examine how the body satisfaction of young women with a healthy BMI is influenced by viewing their bodies in 3D within a virtual environment. By recruiting young women with healthy BMIs, we measured the body satisfaction change of a group that should have the least concerns about their bodies.

The findings of our study revealed that participants' body satisfaction decreased in 14 out of 22 areas measured and increased in seven areas. This indicated a difference in body satisfaction before and after viewing their 3D body scans. Moreover, significant decreases were observed in satisfaction with body build, appearance, waist, and abdomen. These findings align with contemporary beauty norms that idealize thin, slim, and athletic body types for women (Roberts et al. 2022; Vuong et al. 2021). Notably, in the body satisfaction scale, significant changes were observed for 50 percent of general aesthetic features (i.e., body build and appearance). However, only 11 percent of features related to specific body parts (i.e., waist and abdomen) were significantly different. Body build is often defined as the relationship between muscle mass and fat, shaping an athletic or strong appearance (Deurenberg et al. 1999). On the other hand, appearance refers to "the way a person or thing looks to other people" (Cambridge Dictionary n.d.). This indicates that people may care more about how their body is perceived as a whole rather than the specific details of individual body parts. Seeing a 3D body scan can prompt self-reflection similar to viewing oneself from someone else's perspective, emphasizing overall appearance rather than isolated features.

The study findings also showed no significant relationship between BMI and the changes in body satisfaction after viewing the 3D body model. It should be noted that all participants in this study had BMIs within the healthy weight range. This suggests that even young women with healthy BMIs can experience a decrease in satisfaction with features such as waist, abdomen, appearance, and body build. These changes are likely influenced more by societal beauty ideals than by actual BMI (Glauert et al. 2008). Although body satisfaction decreased for several items after viewing the 3D body scans, participants were generally comfortable with the experience, and their comfort levels were not significantly correlated with BMI. This finding raises an important question about whether virtual try-on technologies could inadvertently harm users' self-perception, as users might not be consciously aware of the adverse psychological effects of viewing their 3D body models.

As with any research, this study had limitations, such as using Likert-type scale measurements for body satisfaction. While these scales provide quantifiable data, they lack the depth of insight that qualitative methods, like interviews, could provide. Additionally, with mobile scanning technologies advancing rapidly and producing colorful, photorealistic 3D body models, future research should examine how these realistic scans impact body satisfaction and comfort levels. Despite these limitations, this study makes valuable contributions to the field. It examined changes in body satisfaction with specific body parts, such as the hips and abdomen, and provided a detailed understanding of how 3D body imaging technologies can influence self-perception. Additionally, focusing on young women with healthy BMIs offers a unique perspective, as this group was expected to experience minimal changes in body satisfaction due to 3D body imaging. The decrease in body satisfaction within this group suggests that similar or more significant declines would be likely in other populations, which should be examined in further study.

REFERENCES

Apeagyei, Phoebe R. 2008. Significance of body image among UK female fashion consumers: The cult of size zero, the skinny trend. *International Journal of Fashion Design Technology and Education* 1, Nr. 1 (1. March): 3–11. https://doi.org/10.1080/17543260701867697, https://doi.org/10.1080/17543260701867697.

Arsenyan, J. and A. Mirowska. 2021. Almost human? A comparative case study on the social media presence of virtual influencers. *International Journal of Human-Computer Studies*. https://doi.org/10.1016/j.ijhcs.2021.102694.

Benton, Catherine and Bryan T. Karazsia. 2014. The effect of thin and muscular images on women's body satisfaction. *Body Image* 13 (18. December): 22–27. https://doi.org/10.1016/j.bodyim.2014.11.001, https://doi.org/10.1016/j.bodyim.2014.11.001.

By The Learning Network. 2022. *What Students are Saying about How Social Media Affects Their Body Image*. https://www.nytimes.com/2022/03/31/learning/what-students-are-saying-about-how-social-media-affects-their-body-image.html (Accessed: May 14, 2024).

Calogero, R.M., M. Boroughs and J.K. Thompson. 2007. The impact of western beauty ideals on the lives of women and men: A sociocultural perspective. In V. Swami and A. Furnham (Eds.), *Body Beautiful: Evolutionary and Sociocultural Perspectives* (pp. 259–298). New York: Palgrave Macmillan.

Cambridge Dictionary. n.d. Appearance. *Cambridge Dictionary*. https://dictionary.cambridge.org/dictionary/english/appearance (Accessed: May 15, 2024).

Chattaraman, Veena and Nancy Ann Rudd. 2006. Preferences for aesthetic attributes in clothing as a function of body image, body cathexis and body size. *Clothing and Textiles Research Journal* 24, Nr. 1 (1. January): 46–61. https://doi.org/10.1177/0887302x0602400104, https://doi.org/10.1177/0887302x0602400104.

Cromley, Taya, Stephanie Knatz, Roxanne Rockwell, Dianne Neumark-Sztainer, Mary Story and Kerri Boutelle. 2012. Relationships between body satisfaction and psychological functioning and weight-related cognitions and behaviors in overweight adolescents. *Journal of Adolescent Health* 50, Nr. 6 (10. February): 651–653. https://doi.org/10.1016/j.jadohealth.2011.10.252, https://doi.org/10.1016/j.jadohealth.2011.10.252.

Daanen, H.A.M. and F.B. Ter Haar. 2013. 3D whole body scanners revisited. *Displays* 34, Nr. 4 (28. August): 270–275. https://doi.org/10.1016/j.displa.2013.08.011, https://doi.org/10.1016/j.displa.2013.08.011.

Deng, Fengyi and Xia Jiang. 2023. Effects of human versus virtual human influencers on the appearance anxiety of social media users. *Journal of Retailing and Consumer Services* 71 (March): 103233. https://doi.org/10.1016/j.jretconser.2022.103233.

Deurenberg, P., M. Deurenberg Yap, J. Wang, F.P. Lin and G. Schmidt. 1999. The impact of body build on the relationship between body mass index and percent body fat. *International Journal of Obesity* 23, Nr. 5 (1. May): 537–542. https://doi.org/10.1038/sj.ijo.0800868, https://doi.org/10.1038/sj.ijo.0800868.

Fredrickson, Barbara L. and Tomi-Ann Roberts. 1997. Objectification theory: Toward understanding women's lived experiences and mental health risks. *Psychology of Women Quarterly* 21, Nr. 2 (1. June): 173–206. https://doi.org/10.1111/j.1471-6402.1997.tb00108.x, https://doi.org/10.1111/j.1471-6402.1997.tb00108.x.

Glauert, Rebecca, Gillian Rhodes, Sue Byrne, Bernhard Fink and Karl Grammer. 2008. Body dissatisfaction and the effects of perceptual exposure on body norms and ideals. *International Journal of Eating Disorders* 42, Nr. 5 (29. December): 443–452. https://doi.org/10.1002/eat.20640, https://doi.org/10.1002/eat.20640.

Grogan, Sarah, Simeon Gill, Kathryn Brownbridge, Sarah Kilgariff and Amanda Whalley. 2013. Dress fit and body image: A thematic analysis of women's accounts during and after trying on dresses. *Body Image* 10, Nr. 3 (17. April): 380–388. https://doi.org/10.1016/j.bodyim.2013.03.003, https://doi.org/10.1016/j.bodyim.2013.03.003.

Grogan, Sarah, Simeon Gill, Kathryn Brownbridge, Gillian McChesney, Paula Wren, Jenny Cole, John Darby, Celina Jones and Christopher J. Armitage. 2020. Body mass index and body satisfaction: Does availability of well-fitting clothes matter? *Clothing and Textiles Research Journal* 39, Nr. 4 (1. April): 251–263. https://doi.org/10.1177/0887302x20915528, https://doi.org/10.1177/0887302x20915528.

Grogan, Sarah, Simeon Gill, Kathryn Brownbridge, Dawn Warnock and Christopher J. Armitage. 2015. Women's long-term reactions to whole-body scanning. *Clothing and Textiles Research Journal* 34, Nr. 1 (7. September): 61–73. https://doi.org/10.1177/0887302x15603117, https://doi.org/10.1177/0887302x15603117.

Grogan, Sarah, Emma Storey, Kathryn Brownbridge, Simeon Gill, Clair Templeton, Jayne Gill, Sofia Persson and Christopher J. Armitage. 2020. Whole body scanning as a tool for clothing sizing: Effects on women's body satisfaction. *Journal of the Textile Institute* 111, Nr. 6 (27. September): 862–868. https://doi.org/10.1080/00405000.2019.1668127, https://doi.org/10.1080/00405000.2019.1668127.

Hawkins, Nicole, P. Scott Richards, H. Mac Granley and David M. Stein. 2004. The impact of exposure to the thin-ideal media image on women. *Eating Disorders* 12, Nr. 1 (1. January): 35–50. https://doi.org/10.1080/10640260490267751, https://doi.org/10.1080/10640260490267751.

Idrees, S., S. Gill and G. Vignali. 2023. Mobile 3D body scanning applications: A review of contact-free AI body measuring solutions for apparel. *The Journal of the Textile Institute* 1–12. https://doi.org/10.1080/00405000.2023.2216099.

LaBat, Karen L. and Marilyn R. DeLong. 1990. Body cathexis and satisfaction with fit of apparel. *Clothing and Textiles Research Journal* 8, Nr. 2 (1. January): 43–48. https://doi.org/10.1177/0887302x9000800206, https://doi.org/10.1177/0887302x9000800206.

Lafever, Hayden A., Thomas M. Brinthaupt, Frederick S. Cottle and Mary Beth Asbury. 2019. Changes in body esteem after viewing a 3D full-body representation of oneself. *Journal of Family & Consumer Sciences* 111, Nr. 4 (15. December): 24–34. https://doi.org/10.14307/jfcs111.4.24, https://doi.org/10.14307/jfcs111.4.24.

Loker, Suzanne, Lora Cowie, Susan Ashdown and Van Dyk Lewis. 2004. Female consumers' reactions to body scanning. *Clothing and Textiles Research Journal* 22, Nr. 4 (1. September): 151–160. https://doi.org/10.1177/0887302x0402200401, https://doi.org/10.1177/0887302x0402200401.

Maes, Chelly and Orpha De Lenne. 2022. Filters and fillers: Belgian adolescents' filter use on social media and the acceptance of cosmetic surgery. *Journal of Children and Media* 16, Nr. 4 (25. May): 587–605. https://doi.org/10.1080/17482798.2022.2079696, https://doi.org/10.1080/17482798.2022.2079696.

McComb, Sarah E. and Jennifer S. Mills. 2022. The effect of physical appearance perfectionism and social comparison to thin-, slim-thick-, and fit-ideal Instagram imagery on young women's body image. *Body Image* 40 (27. December): 165–175. https://doi.org/10.1016/j.bodyim.2021.12.003, https://doi.org/10.1016/j.bodyim.2021.12.003.

Mead, George H. 1934. *Mind, Self, and Society*. Chicago: University of Chicago Press.

Möri, Michelle, Federica Mongillo and Andreas Fahr. 2022. Images of bodies in mass and social media and body dissatisfaction: The role of internalization and self-discrepancy. *Frontiers in Psychology* 13 (20. December). https://doi.org/10.3389/fpsyg.2022.1009792, https://doi.org/10.3389/fpsyg.2022.1009792.

Muehlenkamp, Jennifer J. and Renee N. Saris-Baglama. 2002. Self-objectification and its psychological outcomes for college women. *Psychology of Women Quarterly* 26, Nr. 4 (1. December): 371–379. https://doi.org/10.1111/1471-6402.t01-1-00076, https://doi.org/10.1111/1471-6402.t01-1-00076.

The Museum at FIT. 2018. Symposium | Emma McClendon "The body: Fashion and physique" [Video]. *YouTube* (13. June). https://www.youtube.com/watch?v=WBQFq7zw5R4.

The Museum at FIT. n.d. *The Body: Fashion and Physique*. https://www.fitnyc.edu/museum/exhibitions/the-body-fashion-physique.php (Accessed January 29, 2025).

Myers, Taryn A. and Janis H. Crowther. 2007. Sociocultural pressures, thin-ideal internalization, self-objectification, and body dissatisfaction: Could feminist beliefs be a moderating factor? *Body Image* 4, Nr. 3 (20. June): 296–308. https://doi.org/10.1016/j.bodyim.2007.04.001, https://doi.org/10.1016/j.bodyim.2007.04.001.

Neumark-Sztainer, Dianne, Susan J. Paxton, Peter J. Hannan, Jess Haines and Mary Story. 2006. Does body satisfaction matter? Five-year longitudinal associations between body satisfaction and health behaviors in adolescent females and males. *Journal of Adolescent Health* 39, Nr. 2 (1. August): 244–251. https://doi.org/10.1016/j.jadohealth.2005.12.001, https://doi.org/10.1016/j.jadohealth.2005.12.001.

Neyret, Solène, Anna I. Bellido Rivas, Xavi Navarro and Mel Slater. 2020. Which body would you like to have? The impact of embodied perspective on body perception and body evaluation in immersive virtual reality. *Frontiers in Robotics and AI* 7 (18. March). https://doi.org/10.3389/frobt.2020.00031, https://doi.org/10.3389/frobt.2020.00031.

Oosterom, Evrim Buyukaslan, Fatma Baytar and Mona Maher. 2023. Fashion marketing with virtual humans as influencers. In *CRC Press eBooks* (pp. 163–178). (5. April). https://doi.org/10.1201/9781003264958-13, https://doi.org/10.1201/9781003264958-13.

Orbach, Susie. 1978. *Fat is a Feminist Issue*. London: Paddington Press.

Park, Juyeon. 2018. The effect of virtual avatar experience on body image discrepancy, body satisfaction and weight regulation intention. *Cyberpsychology Journal of Psychosocial Research on Cyberspace* 12, Nr. 1 (25. July). https://doi.org/10.5817/cp2018-1-3, https://doi.org/10.5817/cp2018-1-3.

Park, Juyeon and Jennifer Paff Ogle. 2021. How virtual avatar experience interplays with self-concepts: The use of anthropometric 3D body models in the visual stimulation process. *Fashion and Textiles* 8, Nr. 1 (4. July). https://doi.org/10.1186/s40691-021-00257-6, https://doi.org/10.1186/s40691-021-00257-6.

Popescu, Luminița. 2022. Image editing tools for visually appealing self-presentations: Digitally mediated social presence, body dissatisfaction and shame, and ideal beauty standards. *Journal of Research in Gender Studies* 12, Nr. 2: 48. https://doi.org/10.22381/jrgs12220223.

Ridgway, Jessica L. 2017. Before and after avatar exposure. *Clothing and Textiles Research Journal* 36, Nr. 2 (26. December): 91–103. https://doi.org/10.1177/0887302x17749924, https://doi.org/10.1177/0887302x17749924.

Roberts, Savannah R., Anne J. Maheux, Rowan A. Hunt, Brianna A. Ladd and Sophia Choukas-Bradley. 2022. Incorporating social media and muscular ideal internalization into the tripartite influence model of body image: Towards a modern understanding of adolescent girls' body dissatisfaction. *Body Image* 41 (17. March): 239–247. https://doi.org/10.1016/j.bodyim.2022.03.002, https://doi.org/10.1016/j.bodyim.2022.03.002.

Secord, Paul F. and Sidney M. Jourard. 1953. The appraisal of body-cathexis: Body-cathexis and the self. *Journal of Consulting Psychology* 17, Nr. 5 (1. January): 343–347. https://doi.org/10.1037/h0060689, https://doi.org/10.1037/h0060689.

Shim, Soyeon, Antigone Kotsiopulos and Diane S. Knoll. 1991. Body cathexis, clothing attitude, and their relations to clothing and shopping behavior among male consumers. *Clothing and Textiles Research Journal* 9, Nr. 3 (1. March): 35–44. https://doi.org/10.1177/0887302x9100900306, https://doi.org/10.1177/0887302x9100900306.

Shin, Eonyou and Fatma Baytar. 2013. Apparel fit and size concerns and intentions to use virtual try-on. *Clothing and Textiles Research Journal* 32, Nr. 1 (16. December): 20–33. https://doi.org/10.1177/0887302x13515072, https://doi.org/10.1177/0887302x13515072.

Slater, Amy and Marika Tiggemann. 2002. A test of objectification theory in adolescent girls. *Sex Roles* 46, Nr. 9–10 (1. January): 343–349. https://doi.org/10.1023/a:1020232714705, https://doi.org/10.1023/a:1020232714705.

Statista. 2022. *Share of Internet Users in the United States Who Would Be Interested in Creating an Avatar for the Metaverse as of March 2022, by Generation.* https://www.statista.com/statistics/1229349/metaverse-avatar-creation-interest-us-by-generation/ (Accessed: January 29, 2025).

Szambolics, Julia, Sonia Malos and Delia Cristina Balaban. 2023. Adolescents' augmented reality filter usage on social media, developmental process, and well-being. *Media and Communication* 11, Nr. 4 (11. October). https://doi.org/10.17645/mac.v11i4.7016, https://doi.org/10.17645/mac.v11i4.7016.

Tan Ngo, N. 2019. What historical ideals of women's shapes teach us about women's self-perception and body decisions today. *AMA Journal of Ethics* 21, Nr. 10: E879–E901. https://doi.org/10.1001/amajethics.2019.879.

Tiggemann, M. and Clare Lacey. 2009. Shopping for clothes: Body satisfaction, appearance investment, and functions of clothing among female shoppers. *Body Image* 6, Nr. 4: 285–291. https://doi.org/10.1016/j.bodyim.2009.07.003.

Tiggemann, M. and J.E. Lynch. 2001. Body image across the life span in adult women: The role of self-objectification. *Developmental Psychology* 37, Nr. 2: 243–253.

Vandenbosch, Laura, Jasmine Fardouly and Marika Tiggemann. 2022. Social media and body image: Recent trends and future directions. *Current Opinion in Psychology* 45: 101289. https://doi.org/10.1016/j.copsyc.2021.12.002.

Vuong, An T., Hannah K. Jarman, Jo R. Doley and Siân A. McLean. 2021. Social media use and body dissatisfaction in adolescents: The moderating role of thin- and muscular-ideal internalisation. *International Journal of Environmental Research and Public Health* 18, Nr. 24 (15. December): 13222. https://doi.org/10.3390/ijerph182413222, https://doi.org/10.3390/ijerph182413222.

WGSN Insider. 2022. *Trend Q&A: Meta-Realities, 3D Fashion Design and More | WGSN.* https://www.wgsn.com/en/blogs/trend-qa-meta-realities-3d-fashion-design-and-more (Accessed: May 14, 2024).

7 Digital Avatars as Fashioned Bodies

Identity Making and Boundaries

Natalia Särmäkari, Emily Huggard, and Daniel Drak

7.1 INTRODUCTION

The fashion world is one of the key sites for the aesthetics of gendered, politicized, and situated bodies and dress practices (Vänskä et al. 2023; Entwistle 2023). Digitalization of fashion practices has introduced new bodies in fashion: avatars. Avatars can be realistic representations of real humans, or transcend fleshy existence by creating idealized body image extensions in virtual worlds, or entirely different types of human or non-human bodies. Virtual worlds and games offer users an ever-increasing array of options for avatar embodiment and extension of the self (Belk 2013; Won & Davis 2024; Lehtonen et al. 2023; Marder et al. 2019; Ma 2025). This allows users to be anyone or anything they prefer (Shuvalova & Tanaka 2023, 32). In this context, the concept of virtual worlds (also referred to as the 'metaverse') entails digital platforms and games within which users can create and present their identities through appearance in digital form. This includes popular immersive game spaces like *Roblox* and *Fortnite* (Venturini & Columbano 2024, 13–15; Institute of Digital Fashion [IoDF] 2021, 7). Game platforms define the affordances of identity-building possibilities and the nature of social interaction and participation (Drak & Barry 2023, 210).

The online game platform *Roblox's* Metaverse Fashion Trend Report on digital expression within fashion and beauty claims that "[o]ver half (56%) of Gen Z say styling their avatar is more important to them than their physical style" (*Roblox* Corporation 2023). According to digital fashion proponents, digital fashion culture could potentially reshape the body norms and beauty ideals of the fashion industry as people have more freedom to express themselves as avatars in digital spaces who have fewer physical and social restrictions (The Naked Collector and The Fabricant 2024; IoDF 2021; Särmäkari 2021; Park & Chun 2023; Ma 2025). However, self-representation in digital spaces is often an idealized representation of the self, affected by social norms and ideals of real lifeworlds (Zimmerman et al. 2022; IoDF 2021).

 DOI: 10.1201/9781003594482-9

Are avatars as socially restricting as dressed bodies in the physical realm? How are identity and expression limited by programed genderlock (Moncelsi 2021), white habitus in digital spaces (Dietrich 2013), and a general lack of customization options or representation for racialized bodies (Jackson 2020; St Fleur & DeWinter 2021), fat bodies (Grayson 2014), disabled bodies (Nuli 2023), or queerness (Gaudszun & Elmezeny 2023)? We examine fashioned digital avatars of the digital game platform *Roblox* through the lens of situated bodily practice (Entwistle 2023), and the concepts of social belonging, identity-making, boundaries, and vulnerability.

7.2 SITUATED BODILY PRACTICE IN THE SOCIAL VIRTUAL WORLDS

Fashion scholar Joanne Entwistle originally developed the influential theory of situated bodily practice in 2000, basing her argument on Erving Goffman's (1972) account of the body as a property and vehicle of identity between the individual and the social world, defined and managed according to social situations. Individuals feel socially and morally obliged to perform their identities through appearance in accordance with a particular context. Goffman adds to Mary Douglas's idea of two bodies – the physical and social body – whereupon dress forms a link between individual identity and the body and between individual identity and social belonging (Entwistle 2023, 14).

Today, identities are mediated, mediatized, and performed across an array of digital environments, including social media, immersive virtual spaces, and gaming platforms (Marshall 2015; Hudson-Miles 2023; Lehtonen et al. 2023; Won & Davis 2024). These digital spaces serve as sites of identity construction and negotiation, where users create, modify, and perform self-representations in ways that are shaped by media affordances and socio-cultural norms (Rocamora 2017). Media increasingly influences how we construct our personas, bodies, and fashion identities, shaping mediatized versions of our public selves (Marshall 2015, 115). Moreover, digital environments are not merely reflections of the physical world but are social realities of their own, deeply interwoven with our physical realities, culture, and behavior (Schultze 2014; Wolf 2008).

Within these spaces, avatars serve as dynamic, interactive embodiments of the user, operating across social media, virtual worlds, and digital games (Bhardwaj 2023). This interplay between real and digital embodiment has been conceptualized as "real-virtuality" (Shuvalova & Tanaka 2023) or "phygital" experience – a blending of physical and digital embodied worlds (Beaufils & Berland 2022). Avatars function as fashioned bodies, often adorned in digital clothing and accessories that serve as markers of identity, status, and group affiliation (Drak & Barry 2023, 200; Karaminas 2012; Ruberg 2019). These fashioned avatars navigate social norms and identity politics much like physical bodies do in offline contexts, engaging in acts of self-presentation, gender expression, and cultural belonging (Soh et al. 2024). Importantly, digital fashion offers new possibilities for fluid and intersectional identity expression, enabling users to play with race, gender, and body norms in ways that may not be as accessible in the physical world (Mele et al. 2021). The flexibility of representing oneself and acting through an avatar has turned virtual worlds into a social laboratory for exploring identity (Shuvalova & Tanaka 2023; Hudson-Miles 2023).

However, the fluidity of self-representation is not without constraints. Each metaverse community upholds its own set of social roles, values, and norms, which may either reinforce or challenge existing power structures and hegemonic discourses (Won & Davis 2024; Wolf 2008). Furthermore, identity expression is increasingly governed by platform economies and monetization models (Lehtonen et al. 2023; Marder et al. 2019; Nakamura 2013). For example, in *Roblox*, users engage in a situated bodily practice by constructing their avatars through fashion choices that reflect both personal identity and platform-based social dynamics. The customization of avatars – including clothing, hairstyles, and accessories – is shaped by the economic structure of the platform, where access to desirable digital fashion is often contingent on in-game purchases and microtransactions (Lehtonen et al. 2023). As a result, identity formation in platforms like *Roblox* is not only an act of self-expression but also a commercialized process, where platform logic commodifies digital embodiment and social belonging (Nieborg & Poell 2018). The selection of avatar features and clothes is dependent on how much money the players can spend on the platform which is common in free-to-play servitized games.

Ultimately, digital fashion and avatar customization in social virtual worlds reveal the tensions between creative agency, social belonging, and platform capitalism, reflecting broader questions about power, access, and identity politics in the digital age (Rocamora 2022; Nakamura 2013; Bucher & Helmond 2018). In *Roblox*, users engage in a situated bodily practice by operating in communities within and beyond the games, selecting avatar features to construct a visual representation of themselves.

7.3 *ROBLOX* AS A CONTEXT

Roblox is a social game platform, highly popular among youth (Backlinko Team 2025). Recently, fashion companies such as Louis Vuitton and Gucci have capitalized on the platform's marketing and monetary potential by creating branded digital wearables, phygital immersive experiences merging physical and digital consumer engagement (Venturini & Columbano 2024). Self-expression is a core facet of *Roblox* with millions of active users updating and customizing their avatars daily (McDowell 2021; Wootton & Bronstein 2023). The *Roblox* avatars, though sometimes designed to resemble their users, typically embody a more stylized and block-like aesthetic, differing from the hyper realistic avatars seen in platforms like *Fortnite* or Meta's *Horizon Worlds*. Players engage in fashioning their digital selves through advanced 3D-modeled outfits, accessories, and animations, which contribute to an evolving virtual fashion ecosystem.

Customization in *Roblox* is driven by user-generated content (UGC), where players not only modify existing assets but also create entirely new clothing items, accessories, and environments. The platform provides development tools such as Roblox Studio and the Avatar Marketplace, enabling users to design, sell, and trade digital fashion assets (Fernández-Vara 2015; Drak & Barry 2023). This democratization of content creation allows for a decentralized fashion economy, where young designers can monetize their work and gain recognition within the community. *Roblox* has an internal virtual currency called 'Robux', required to purchase most avatar garments and customization assets. Unlike in traditional retail, where consumers buy finished products, in *Roblox*, users often engage in tiered monetization models, where fashion items may be locked behind limited-edition drops, collaborations, or in-game achievements (Lehtonen et al. 2023).

7.4 METHODS

Our study focuses on the case of *Roblox* and is guided by the question of whether avatars are as socially restricting as dressed bodies in the physical realm. We triangulate a variety of research material consisting of documents produced by *Roblox*, public conversations on the r/Roblox Reddit and various *Roblox* subreddits, and our own observations whilst playing *Roblox*. We analyzed 15 subreddits that focused on appearance, body, and outfits in *Roblox* between February and May 2024. We created our own avatars and played popular fashion-related games including *Fashion Show*, *Fashion Famous*, *Dress To Impress*, and *Passion4fashion*. Our goal was to investigate how users interact with and shape their avatars on the platform and understand the social and embodied aspects of identity-making. We also analyzed the reports highlighting self-expression published by *Roblox* (Roblox Corporation 2023) and Roblox & Parsons School of Design (Roblox Corporation 2022). In fact, these reports – arguing for the high importance of avatars to Gen Z – were the starting point of our research question. If avatars are such an essential part of their identity experience, do similar social boundaries restricting our appearances and embodied practices in the physical world also apply to the social virtual world?

We apply textual game analysis to study the meanings within and around the game platform (Fernández-Vara 2015, 6). By textual analysis, we refer to a methodology that seeks an understanding of "language, symbols, and/or pictures present in texts to gain information regarding how people make sense of and communicate life and life experiences" (Hawkins 2017, 1754). The text of the game is inevitably produced by the players who interact and are in dialogue with the game and each other (Fernández-Vara 2015, 7–8), and in the case of *Roblox*, actively shape and even create the games within the platform. We recognize that the text of the game is co-produced by players, who engage in creation, customization, performance, and dialogue with the game and each other. *Roblox*'s participatory structure requires analysis beyond the formal elements of the platform (e.g., avatar customization options, rules, and affordances) but also the emergent meanings produced through play, interaction, and discourse within the community.

7.5 FINDINGS

Drawing from our observation and qualitative analysis of user interactions, game mechanics, and self-expression practices within the *Roblox* platform and its surrounding communities in Reddit, we generated three major themes: community sentiment and social belonging, identity-making and vulnerability, and sublime phygital embodiment. These themes illuminate how digital dress operates as both a mode of identity construction and a reflection of broader social dynamics in virtual spaces.

7.5.1 Community Sentiment and Social Belonging

Humans are social beings who wear clothes and adorn their bodies in and for their communities. Though dress in the physical realm has creative and expressive qualities to some extent, Entwistle (2023, 7–12) argues that identity is mainly managed through dress in more mundane ways as well as the context and social pressure that defines the boundaries of the "normal" and "appropriate". Particular contexts constrain what one chooses to wear.

Dressing avatars based on community sentiment and social context was evident in our analysis. In *Roblox*, users frequently interact with others through social media platform Reddit (r/Roblox), gameplay, and virtual events. These interactions influence users' decisions regarding their avatar's appearance, as they may seek to conform to or differentiate themselves from social norms and peer preferences. In the Reddit post "In what way do you guys like to dress your avatars?" one user commented, "Basic, I like to have no impression on other people so they won't assume anything lol" (Reddit user 2021).

In another Reddit post titled "Does my outfit look like a slender or emo?", fears of being labeled or placed into a specific subcultural group – such as a "Slender" (*Roblox* players who follow a goth-punk style with their characters, often featuring a thin and extra-tall body) or "Emo" (a term derived from a niche music genre that now describes a style characterized by skinny jeans and band T-shirts) – demonstrate the need for belonging and alignment with social norms (Reddit user 2023a).

Various discussions on Reddit also expressed the fear of being judged on the platform, with users noting that people often bully others for how they dress in *Roblox* more than in "real life". In the Reddit post titled "It's kinda dumb to care about people's avatars," users note:

> This is a kid's game where people can dress how they want, yet people get insulted for dressing how they want. People get bashed for wearing specific hats, and it's really dumb. If you wear the bear mask or holiday crown, you're bashed and called an ODer [online dater]. Wearing 1.0 body parts? You're bashed and called fat. I could go on and on, but I won't. Why would you care how other people look on *Roblox* of all places?
>
> (Reddit user 2020)

This contradicts the imaginaries of *Roblox*, which highlight the explorative nature of digital spaces as sites for experimentation with identities through fashion "while engaging in worldmaking and building communities" (Roblox Corporation 2022, 23) and according to which their infrastructure "makes self-expression opportunities truly limitless" (ibid., 2). In the games *Dress To Impress* and *Passion4fashion*, self-expression is encouraged as users dress their characters in outfits, choosing individual pieces and then walking a virtual runway. More broadly on the platform, the

Roblox currency (Robux) influences the outcome of the avatars and mimics physical societal hierarchies. *Roblox* players can use Robux to purchase a vast range of items for their avatar from third-party apps. These vary widely in cost – from $1 to accessories that can cost up to $1000. Social pressures and social hierarchies are evident in the existence of "rare" items (similar to luxury goods) which creates inequality and competition through manufactured scarcity. Users are often pressured to keep up with peers' avatars and to buy gifts for others. The dependency on the economic capital of players sets boundaries to the types of items that can be purchased and hypothetically constructs similar hierarchies among fashioned bodies in games as there are in the physical world.

In each game, players can choose from a variety of garment options to earn points and stars, climbing leaderboards based on adherence to certain themes. While playing the games, experimenting with one's identity is of prime importance, as opposed to expressing actual self-identity. The chosen clothing items are extravagant, colorful, and conspicuous, encouraging dress and fashion to be playful, similar to the runway in the physical world. In *Dress To Impress*, players rank other players' styles and fashion choices as part of a community-based fashion show. In a Reddit thread titled "*Roblox* Dress to Impress New Themes are TOXIC", it is mentioned that the platform has "two very hyper-feminine/masculine body options with no variations for other heights or weights", illustrating the restrictions in projecting actual or idealized identity (Reddit user 2024).

7.5.2 Identity-Making, Boundaries, and Vulnerability

The communities within *Roblox* and on platforms such as Reddit incorporate and build social structures of their own, and may be considered to form what media scholar Lauren Berlant (2008) calls 'intimate publics', referring to communities that emotionally unite social groups that do not know each other in real life and would not have the feeling of social belonging without the connecting media. Both avatars and our digital possessions alike could be seen as extensions of our embodied selves, shaping our identities in real life (Belk 2013) and connecting to others. *Roblox* avatars enable users to perform and experiment with different aspects of their identity within the virtual environment which may or may not be aligned with their dressed body in the physical realm. Virtual worlds have been spaces for such identity exploration not constrained by the boundaries of the physical world, including gender and ability exploration (Huh & Williams 2010). However, virtual worlds and games have also proven to perpetuate biases and narrow beauty ideals and social norms (Lehtonen et al. 2023).

Entwistle's framework emphasizes the boundaries of fashion and bodily practices, wherein individuals learn to live in their bodies and construct their identity through their appearance. As Entwistle (2023, 8) writes: "Wearing the right clothes and looking our best, we feel at ease with our bodies, and the opposite is equally true: turning up for a situation inappropriately dressed, we feel awkward, out of place and vulnerable. In this respect, dress is an intimate experience of the body and a public presentation of it." In the virtual world, dress is a boundary between the self and the other, the private and the public. Numerous Reddit discussions over past years regard the questions on the extent to which players can use avatars that do not align with their gender or racial identity. These discussions highlight the complexities of virtual identity formation and revolve around the cultural and ethical implications of digital self-representation. A *Roblox* community member asks: 'Is representing "myself" with a black avatar racist?' (Reddit user 2023c).

The question of whether it is offensive to create avatars outside one's own racial identity reflects larger debates on cultural appropriation versus appreciation in digital fashion. Who has the right to embody which identities, and under what circumstances? While digital spaces offer new possibilities for self-representation and exploration, they also reproduce real-world power dynamics that dictate who gets to experiment with identity freely and who faces scrutiny for doing so. For instance, Tosaya (2022) discusses the notion of "digital blackface" addresses a variety of internet-age actions that reflect a type of minstrelsy, oftentimes referring to how non-Black users might depict Black people in digital spaces via avatars or characters. Understanding such tensions is crucial to examining how race, fashion, and digital embodiment intersect in virtual worlds.

Similarly, digital dress provides opportunities for experimentation and identity expression, yet players often face criticism when they do not conform to social norms. Digital dress provides opportunities for experimentation and identity expression, yet players often face criticism when they do not conform to social norms. For example, in the r/RobloxAvatars subreddit, many debates center on the presence of "furry" and "femboy" avatars. This is evident in discussions such as one initiated by u/Majestic-Office-4942, who posted a meme stating, "This subreddit would be better if people talked about the good and creative avatars, not furry and/or femboy avatars" (Reddit user 2023b). Thus, in *Roblox*, players negotiate the presentation of their avatar's identities within various contexts that might similarly empower and disavow their choices. In this sense, choices surrounding identity vis-a-vis the avatar may be vast, though various socially constructed boundaries seem to influence how such choices are perceived.

7.5.3 Sublime Phygital Embodiment

The avatar bodies are situated in the social realities of virtual communities, entangled with our physical social existence (Shuvalova & Tanaka 2023; Beaufils & Berland 2022). Dressed digital bodies and fleshy bodies merge into a phygital existence, where physical and digital boundaries are blurred, and the identity becomes plural. However, such enmeshment might also overwhelm and disguise the social realities, including the "image-driven superficiality and overconsumption of late capitalist society" in the form of the technological sublime (Hudson-Miles 2023; Jameson 1991). While the sublime experience has been associated with natural landscapes, in Jameson's account, technology can also provoke "bewilderment and disorientation which is, momentarily, sublime" (Jameson 1991, 50). The term phygital embodiment refers to the entanglement of physical and digital forms of existence, where virtual bodies serve as extensions, reflections, or divergences of fleshy selves.

Within *Roblox* and similar spaces that prioritize avatar design, players experiment with various aesthetic and identity markers. Each choice, as it pertains to dress, style, or the body, reflects the influence or interpretation of complex social, cultural, political, and economic systems. We propose the concept of sublime phygital embodiment in which the virtual and physical lifeworlds collide, unveiling the processes of identity-making, as well as disguised social boundaries and vulnerability in idealized cyberspaces. We are incapable of comprehending the complexity of our socio-technological world with a global "multinational and decentralized communication network in which we find ourselves caught as individual subjects" (Jameson 1991, 44). Our observations in the *Roblox* communities align with earlier studies on social norms and body image in virtual worlds, suggesting that identities are adjusted similarly to "expectations of acquisition of social, financial, or amusement benefits in virtual worlds" (Won & Davis 2024, 887).

We recontextualize here the concept of the sublime, traditionally associated with awe and transcendence, as the sensation of exceeding physical limitations through digital identity formation. For some users, creating avatars outside their offline identity fosters a sense of experimentation and boundary dissolution. Others, however, navigate anxieties over appropriation, authenticity, and digital ethics. This tension embodies the sublime paradox where embodiment in virtual space is simultaneously expansive and constrained by socio-cultural norms.

7.6 CONCLUSIONS

Digital embodiment presents a complex intersection of self-expression, platform affordances, and social norms. *Roblox*, as a social and customizable platform, enables users to craft avatars that may diverge significantly from their offline bodies and identities. However, these acts of digital identity construction are not purely individual choices; they are influenced by broader cultural, technological, and discursive structures (cf. Entwistle 2023). In our ongoing research, we studied *Roblox* as the site for building situated and embodied digital identities. The culture of digital fashion is still in development, and it is important to examine how social worlds evolve over time. We argue that the promise of limitless self-expression and the increasing importance of digital identities should be examined in relation to situated bodily practices and the disguising or idealizing processes of sublime phygital embodiment.

We propose that (1) the avatar is a microcosm of complex systems, (2) seemingly increasing "limitless" choices and creativity are shaped by structural constraints, and (3) identity negotiation occurs within the tension between self-expression and socio-cultural influence. Looking at the choices of avatar identity in relation to the sublime, we note how the vastness of diverse systems (social, cultural, political, and economic) and their interpretations are all interconnected in avatar choices and evolving popular styles. These systems could be considered as boundaries while at the same time offering a space for expressing political, economic, and cultural identities that are under constant negotiation between belonging and differentiation. Our findings show that virtual environments like *Roblox* and its communities replicate many of the same pressures, boundaries, and inequalities found in the physical world. Avatar construction and digital dress, far from being merely aesthetic or playful, emerge as a site of identity-making where users navigate experimentation, acceptance, and vulnerability.

The seemingly 'limitless' capacity for user-generated content is critical to *Roblox*'s success. It counters the often stagnant and limited range of identities in many popular games by affording individuals the ability to create and customize their own content. Typically, free-to-play mobile games balance the provision of creative freedom with controlled engagement between the player and the game due to the monetization practices of gaming (Nieborg & Poell 2018; Lehtonen et al. 2023). In the case of *Roblox*, building avatar identities, creating assets within the games, and even developing new games is one of the core monetization methods. This approach helps address the lack of representation of identities in avatar design specifically when these choices are determined by the players rather than by a centralized game studio. However, despite the availability of near-limitless choices and creative possibilities concerning avatar design and identity, players are still subject to policing regarding their fashion and body choices.

We studied a snapshot of the fashion- and identity-related conversations about *Roblox* on Reddit and played a handful of games on the platform for a short period of time. Further research is required to deepen the understanding of virtual identity-making through avatars and community discussions. Future studies should involve an extended period of gameplay, a broader analysis of online conversations on platforms such as Reddit and Discord, as well as empirical and even ethnographic studies on players in real life. Furthermore, an ethical assessment of the logic behind the *Roblox* platform in relation to player experience is of high importance, as many of the players and creators are underaged.

REFERENCES

Backlinko Team. 2025. "Roblox User and Growth Stats You Need to Know." *Backlinko*, January 17. https://backlinko.com/roblox-users#roblox-monthly-active-users. Accessed March 30, 2025.

Beaufils, Kevin, and Alexis Berland. 2022. "Avatar Embodiment: From Cognitive Self-Representation to Digital Body Ownership." *Hybrid: Revue des arts et médiations humaines* 9. Translated by Armelle Chrétien. https://doi.org/10.4000/hybrid.2664.

Belk, Russell W. 2013. "Extended Self in a Digital World." *Journal of Consumer Research* 40: 477–501. https://doi.org/10.1086/671052.

Berlant, Lauren. 2008. *The Female Complaint: The Unfinished Business of Sentimentality in American Culture*. Durham: Duke University Press.

Bhardwaj, Pratham. 2023. "The Metaverse and Social Interaction: Building Virtual Communities." *Medium*. https://medium.com/@prathambhardwaj38/the-metaverse-and-social-interaction-building-virtual-communities-9c62b911c7de. Accessed March 30, 2025.

Bucher, Taina, and Anne Helmond. 2018. "The Affordances of Social Media Platforms." In *The SAGE Handbook of Social Media*, edited by Jean Burgess, ThomasPoell, and Alice Marwick, 233–253. London and New York: SAGE Publications Ltd.

Dietrich, David R. 2013. "Avatars of Whiteness: Racial Expression in Video Game Characters." *Social Inquiry* 83 (1): 82–105. https://doi.org/10.1111/soin.12001.

Drak, Daniel, and Ben Barry. 2023. "Modding Masculinities: Video Game Glitches and Transcending Gendered Dress." *Critical Studies in Men's Fashion* 10 (2): 197–215. https://doi.org/10.1386/Csmf_00076_1.

Entwistle, Joanne. 2023. *The Fashioned Body: Fashion, Dress and Modern Social Theory*. 3rd ed. Cambridge: Polity.

Fernández-Vara, Clara. 2015. *Introduction to Game Analysis*. New York and London: Routledge.

Gaudszun, Isabell Charlott, and Ahmed Elmezeny. 2023. "For the Girls, Gays, and Theys: LGBTQ+ Stakeholder Communication and Alignment of Video Game Brands." *Frontiers in Communication* 8: 1167710. https://doi.org/10.3389/fcomm.2023.1167710.

Goffman, Erving. 1972. *The Presentation of Self in Everyday Life*. London: Penguin.

Grayson, Nathan. 2014. "The Challenges of Letting You be a Fat Video Game Hero." *Kotaku*. https://kotaku.com/the-challenges-of-letting-you-be-a-fat-video-game-hero–1637837650. Accessed March 30, 2025.

Hawkins, Jennifer Morey. 2017. "Textual Analysis." In *The SAGE Encyclopedia of Communication Research Methods*, edited by Mike Allen, 1754–1756. Thousand Oaks, CA: SAGE Publications, Inc. https://doi.org/10.4135/9781483381411.

Hudson-Miles, Richard. 2023. "Elevate the Human Race, Putting Makeup on My Face." *Vestoj*. https://vestoj.com/elevate-the-human-race-putting-makeup-on-my-face-ikon-1-and-the-technological-sublime/. Accessed January 23, 2024.

Huh, Searle, and Dmitri Williams. 2010. "Dude Looks Like a Lady: Gender Swapping in an Online Game." In *Online Worlds: Convergence of the Real and the Virtual*, 161–174. London: Springer. https://doi.org/10.1007/978-1-84882-825-4_13

Institute of Digital Fashion. 2021. "My Self, My Avatar, My Identity: Diversity and Inclusivity within Virtual Worlds." *IoDF White Paper*. https://docsend.com/view/wem8e7ppe-7gr4mrk. Accessed April 23, 2024.

Jackson, Gita. 2020. "Black 'The Sims 4' Players are Changing One of the World's Biggest Games." *Vice*, August 12. https://www.vice.com/en/article/4ay5a9/black-the-sims-4-players-are-changing-one-of-the-worlds-biggest-games. Accessed 30 March 2025.

Jameson, Fredric. 1991. *Postmodernism, or the Cultural Logic of Late Capitalism*. Durham: Duke University Press.

Karaminas, Vicki. 2012. "Image: Fashionscapes – Notes Toward an Understanding of Media Technologies and Their Impact on Contemporary Fashion Imagery." In *Fashion and Art*, edited by Adam Geczy, and Vicki Karaminas, 177–187. Oxford: Berg Publishers.

Lehtonen, Miikka J., J. Tuomas Harviainen, and Annakaisa Kultima. 2023. "How Monetization Mechanisms in Mobile Games Influence Consumers' Identity Extensions." *Service Business* 17: 113–136. https://doi.org/10.1007/s11628-022-00518-4

Ma, Jin Joo. 2025. "Restructuring Situated Bodily Practices: Understanding the Relationship of Dress, Body, and Identity in Virtual Dressing Practice." *Fashion Practice* 17 (1): 27–49. https://doi.org/10.1080/17569370.2024.2427330.

Marder, Ben, David Gattig, Emily Collins, Leyland Pitt, Jan Kietzmann, and Antonia Erz. 2019. "The Avatar's New Clothes: Understanding Why Players Purchase Non-Functional Items in Free-to-Play Games." *Computers in Human Behavior* 91: 72–83. https://doi.org/10.1016/j.chb.2018.09.006.

Marshall, P. David. 2015. "Monitoring Persona: Mediatized Identity and the Edited Public Self." *FRAME* 28: 115–133. https://researchoutput.csu.edu.au/ws/portalfiles/portal/266783271/214498958_Pubished_article.pdf

McDowell, Maghan. 2021. "Shaping Online Avatars: Why Our Digital Identities Differ." *Vogue Business*, October 19. https://www.voguebusiness.com/technology/shaping-online-avatars-why-our-digital-identities-differ. Accessed March 30, 2025.

Mele, Christina, Tiziana Russo-Spena, Marco Tregua, and Christina Catarina Amitrano. 2021. "The Millennial Customer Journey: A Phygital Mapping of Emotional, Behavioural, and Social Experiences." *Journal of Consumer Marketing* 38 (4): 420–433. https://doi.org/10.1108/JCM-03-2020-3701

Moncelsi, Katie. 2021. "Difference of Congruency in Avatar Gender Selection on Motivation to Play Massively Multiplayer Online Role-Playing Games." PhD diss., Capella University.

Nakamura, Lisa. 2013. *Cybertypes: Race, Ethnicity, and Identity on the Internet.* New York: Routledge.
The Naked Collector and The Fabricant. 2024. "Into the Deep: Digital Fashion on Chain the Past Five Years & What Is Coming Next." *Report.* https://drive.google.com/file/d/1rAKglECKGFxyDTKk37M0Ay6IIfJ-b1Vb/view. Accessed January 23, 2024.
Nieborg, David B., and Thomas Poell. 2018. "The Platformization of Cultural Production: Theorizing the Contingent Cultural Commodity." *New Media & Society* 20 (11): 4275–4292. https://doi.org/10.1177/1461444818769694
Nuli, Sangita. 2023. "The Sims 4 Needs Disability Representation More Than It Did Infants." *GameRant*, February 14. https://gamerant.com/the-sims-4-disability-representation-necessary-infants-less-important/. Accessed April 25, 2024.
Park, Juha, and Jaehoon Chun. 2023. "Evolution of Fashion as Play in the Digital Space." *Fashion Practice* 15 (2): 256–278. https://doi.org/10.1080/17569370.2022.2149837.
Reddit user. 2020. "It's Kinda Dumb to Care About People's Avatars." *Reddit.* https://www.reddit.com/r/roblox/comments/iylwob/its_kinda_dumb_to_care_about_peoples_avatars/. Accessed March 30, 2025.
Reddit user. 2021. "In What Way Do You Guys Like to Dress Your Avatars?" *Reddit.* https://www.reddit.com/r/roblox/comments/ps13t9/in_what_way_do_you_guys_like_to_dress_your_avatars/. Accessed March 30, 2025.
Reddit user. 2023a. "Does My Outfit Look Like a Slender or Emo?" Reddit. https://www.reddit.com/r/RobloxAvatars/comments/15ufy90/does_my_outfit_look_like_a_slender_or_emo/. Accessed March 30, 2025.
Reddit user. 2023b. "It Has to Be Said." *Reddit.* https://www.reddit.com/r/RobloxAvatars/comments/187ifhu/it_has_to_be_said. Accessed March 30, 2025.
Reddit user. 2023c. "Is Representing "Myself" with a Black Avatar Racist?" [Online Forum Post]. *Reddit.* www.reddit.com/r/RoyaleHigh_Roblox/comments/16igaef/is_representing_myself_with_a_black_avatar_racist/. Accessed March 30, 2025.
Reddit user. 2024. "Roblox Dress to Impress New Themes Are Toxic." *Reddit.* https://www.reddit.com/r/sinjindrowning/comments/199d45v/roblox_dress_to_impress_new_themes_are_toxic/
Roblox Corporation. 2022. *2022 Metaverse Fashion Trend Report.* https://corp.roblox.com/pdf/2022-metaverse-fashion-trends-report. Accessed March 30, 2025.
Roblox Corporation. 2023. "Digital Expression, Fashion & Beauty Trends Report: How Gen Z Express Themselves in Immersive Spaces." *Roblox.* https://cdn.buttercms.com/D5cYYxLSQuvPzIrrP7kq. Accessed March 30, 2025.
Rocamora, Agnès. 2017. "Mediatization and Digital Media in the Field of Fashion." *Fashion Theory* 21 (5): 505–522. https://doi.org/10.1080/1362704X.2016.1173349.
Rocamora, Agnès. 2022. "The Datafication and Quantification of Fashion: The Case of Fashion Influencers." *Fashion Theory* 26 (7): 1109–1133. https://doi.org/10.1080/1362704X.2022.2048527
Ruberg, Bo. 2019. *Video Games Have Always Been Queer.* New York: New York University Press.
Särmäkari, Natalia. 2021. "Digital 3D Fashion Designers: Cases of Atacac and The Fabricant." *Fashion Theory* 27 (1): 85–114. https://doi.org/10.1080/1362704X.2021.1981657.
Schultze, Ulrike. 2014. "Performing Embodied Identity in Virtual Worlds." *European Journal of Information Systems* 23 (1): 84–95. https://doi.org/10.1057/ejis.2012.52
Shuvalova, Mariia, and Shogo Tanaka. 2023. "Self-Identity in a Virtual Space: Consideration from an Embodied Perspective." *Civilization* 31: 27–37. https://www.u-tokai.ac.jp/uploads/2023/04/b4e1ad449654b5a3418ece6dbd92ef4c.pdf
Soh, Serena, Sanaz Talaifar, and Gabriella M. Harari. 2024. "Identity Development in the Digital Context." *Social and Personality Psychology Compass* 18 (2): 1–22. https://doi.org/10.1111/spc3.12940

St Fleur, Abdah, and Jennifer De Winter. 2021. "'Unfiltered and True to Itself': How Content Creators Represent the Black Community in The Sims 4." *American Journal of Play* 13 (2–3): 297–319. https://files.eric.ed.gov/fulltext/EJ1333256.pdf

Tosaya, Lando C. 2022. "The Rise of Digital Misogynoir-Face: Exploitation of Stereotypes for Entertainment and Profit." Master's thesis, University of Washington. https://digital.lib.washington.edu/server/api/core/bitstreams/047b48ce-0c1c-4131-a27f-5c7cbdab1898/content

Vänskä, Annamari, Jenni Hokka, and Natalia Särmäkari. 2023. *Intimacy: Embodied Knowledge, Creative Work and Digitalization in Contemporary Finnish Fashion*. Espoo: Aalto Arts Books.

Venturini, Alice, and Martina Columbano. 2024. "'Fashioning' the Metaverse: A Qualitative Study on Consumers' Value and Perceptions of Digital Fashion in Virtual Worlds." *Journal of Global Fashion Marketing* 15 (1): 6–22. https://doi.org/10.1080/20932685.2023.2234918.

Wolf, Mark J. P. 2008. *The Video Game Explosion: A History from Pong to PlayStation and Beyond*. Westport: Greenwood Publishing Group.

Won, Andrea Stevenson, and Donna Z. Davis. 2024. "Your Money or Your Data: Avatar Embodiment Options in the Identity Economy." *Convergence: The International Journal of Research into New Media Technologies* 30 (2): 882–903. https://doi.org/10.1177/13548565231200187

Wootton, Christina, and Manuel Bronstein. 2023. "2023 Digital Expression, Fashion & Beauty Trends: How Gen Z Express Themselves in Immersive Spaces." *Report by Roblox*. https://blog.roblox.com/wp-content/uploads/2023/11/2023-Roblox-Digital-Expressions-Report.pdf?utm_source=blog&utm_medium=pdf+download&utm_campaign=digital+expressions. Accessed January 30, 2024.

Zimmermann, Daniel, Anne Wehler, and Kai Kaspar. 2022. "Self-Representation through Avatars in Digital Environments." *Current Psychology* 42: 21775–21789. https://doi.org/10.1007/s12144-022-03232-6

8 A Pedagogic Approach to Digital Pattern Cutting

Carolyn Hardaker, Eleanor Bandey, and Karen Deacon

8.1 INTRODUCTION

The fashion industry is undergoing a significant shift toward digital transformation, driven by advancements in technology and an increasing focus on sustainability. This transformation, which accelerated during the COVID-19 pandemic, underscores the urgency of integrating digital methods across the industry's value chain (Gonzalo et al., 2019). One key area where this change is evident is in the adoption of 3D digital pattern cutting, a practice that has reshaped traditional approaches to garment design and production.

Historically, pattern cutting relied on physical tools, such as paper blocks and calico toiles, to translate 2D patterns into 3D garments. However, with the rise of powerful and accessible 3D digital prototyping software, this process is evolving. Increasingly, fashion education programs are recognising the importance of preparing students to use these tools to meet the demands of a digitally-driven industry.

This paper examines how De Montfort University's Fashion Design curriculum has responded to this need by adopting a blended approach to pattern cutting instruction. Combining analogue and digital methods, the program integrates 3D digital technologies into its teaching practices while adhering to core pedagogic principles for practical disciplines.

Using a library of digitised 2D pattern blocks and virtual avatars, the curriculum enables students to conduct pattern cutting exercises in a digital environment. This approach not only enhances the learning experience but also supports sustainability goals by reducing the reliance on physical materials, such as calico.

To evaluate the impact of this approach, a mixed-methods survey was conducted to gather student feedback. The findings reveal how students have incorporated digital methods into their workflows, developed green skills, and engaged with broader sustainability practices. This paper explores these outcomes and identifies best practices for integrating digital pattern cutting into fashion education programs, offering insights into the opportunities and challenges of blending analogue and digital methods.

8.2 PATTERN CUTTING AND THE PEDAGOGIC CONTEXT

8.2.1 Features of the Garment Pattern Cutting Process

Garment pattern cutting is a practical discipline that requires the learner to acquire knowledge and analogue skills to translate an initial sketch into a set of 2D flat

DOI: 10.1201/9781003594482-10

patterns typically using pattern blocks as templates. The consideration of the 3D body form, fabric characteristics, and pattern shape are important parts of the process. Aldrich (1996, 5) states, 'Understanding the relationship between fabric, form, and pattern shape is one of the most important skills a designer has to acquire.'

These variables contribute to the pattern cutting process being an iterative one, with the design and fit being assessed on the 3D body form through the construction of a calico toile or sample. Patterns are adjusted and refined in 3D and 2D and the toile remade until the design is deemed ready to be made in the desired fabric, with further refinements possible to achieve the required aesthetic qualities.

Almond (2016) reiterates this complexity and states, 'The pattern cutter uses skills, knowledge, experience, and judgment, often with untested and unfamiliar combinations of elements that require intellectual, practical and tacit knowledge and synthesis to interpret and resolve in a feasible manner.'

Digital pattern design software applications to replicate 2D pattern cutting processes were developed in the 1970s to speed up the pattern development process and ensure pattern accuracy and measurement precision (Taylor, 1990). Recent developments in computer graphics, computation power, and software have seen the rapid increase in the use of 3D digital pattern cutting systems. These systems integrate with a 2D digital pattern cutting system, enabling the 2D pieces to be 'stitched' together and draped over a mannequin. The garment is visualised accurately in real time from a choice of hundreds of fabric properties, as seen in Figure 8.1.

These accurate visualisations can speed up the design development process and reduce the number of physical samples required. Today they are established across the fashion value chain incorporating a vast array of pattern cutting functionality.

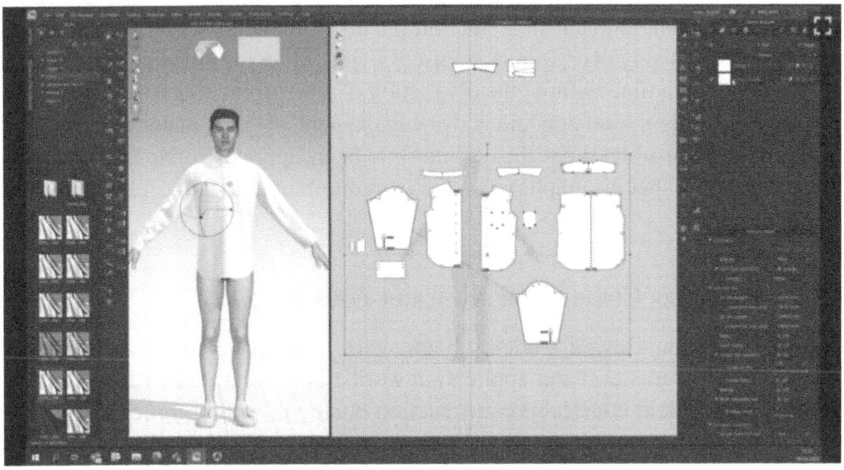

FIGURE 8.1 Clo3D screenshot showing the 2D and 3D workspace and software interface.

8.2.2 Pedagogic Approaches

There are several theories associated with the pedagogy of practical disciplines such as pattern cutting. These include 'constructivism,' where the student transforms from a passive recipient of information to an active participant in the learning process (Hamer and Van Rossum, 2010). Pattern cutting is typically taught by a tutor demonstrating a method or technique with the learner replicating this within the pattern cutting studio.

This studio setting echoes the idea of 'connectivism' and the concept of 'a community of practice,' which emphasises learning as a social process (Wenger, 2010). Working in a community relates to Vygotsky (1978), who considers the development of autonomy in the Zone of Proximal Development. This captures the concept of independent learning and how students develop this ability through use of language and discussions within a community of tutors and fellow students to develop ideas. Having immediate access to tutors and peers provides a learner with the confidence to undertake the set tasks and experiment with their own ideas.

The educational technologist Selwyn (2011, 89) notes the relevance of these theoretical frameworks, developed originally for analogue practice when considering digital teaching and learning. He notes that 'situated learning relates to learning through active participation with an individual involved in the co-construction of knowledge with more able peers and, importantly being held to account for the competence with which they perform.' Selwyn also states that 'this learning is ideally situated in real life environments where the practice or activity takes place.' Taber (2015) comments further the importance of scaffolding facets of traditional studio and workshop design pedagogy. He notes that information exchange, a community of practice, peer-to-peer learning in the digital environment are vital.

8.3 INTEGRATING 3D DIGITAL DESIGN METHODS

In 2022, De Montfort University implemented a block teaching system to enhance student engagement and learning outcomes. This academic structure delivers a unit of learning intensively over a short, focused period, allowing students to concentrate on one subject at a time without the distraction of concurrent courses. This approach encourages deeper immersion and active engagement with the material.

Figure 8.2 illustrates how this system has been applied to the Fashion Design curriculum, with specific emphasis on the integration of analogue and digital pattern cutting methods.

8.3.1 First-Year Curriculum: Analogue Foundations

First-year Fashion Design students begin with analogue pattern cutting, taught through skills acquisition and application workshops integrated with design, illustration, and garment construction instruction during their first three blocks. These sessions take place in a studio environment, with students observing tutor demonstrations via a visualiser and projector. Group sizes are typically kept small, with a 20:1 student-to-tutor ratio, as shown in Figure 8.3.

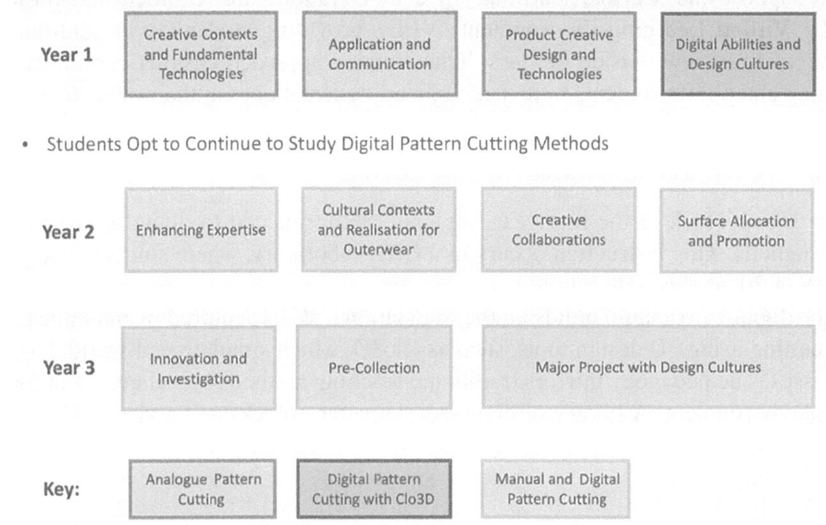

FIGURE 8.2 Fashion design curriculum.

FIGURE 8.3 Analogue pattern cutting class in a studio.

To support and scaffold learning, all demonstrations are recorded and hosted on the Virtual Learning Environment (VLE), providing students with additional resources to review outside of class. This blended approach ensures continuity in learning and allows students to revisit challenging techniques at their own pace.

8.3.2 Transition to Digital Pattern Cutting

In the fourth block of the first year, students are introduced to digital pattern cutting methods. This instruction occurs in a CAD laboratory, where students learn in groups of 20, as shown in Figure 8.4.

The digital curriculum builds on the foundational skills acquired in analogue pattern cutting using 3D design tools, such as Clo3D, which simulate real-world design processes. The pedagogy mirrors traditional teaching methods but adapts them for a digital environment. A library of digitised 2D pattern blocks and a virtual UK size 12 female avatar are used as teaching tools. Students engage in exercises that include the following:

- Draping 2D patterns on the 3D avatar
- Assigning calico fabric properties to assess fit and proportions on screen

These enable students to explore the principles of digital pattern cutting in a manner aligned with their earlier manual experiences. Throughout these sessions, students receive guidance from tutors and peers, ensuring a collaborative and supportive

FIGURE 8.4 Digital pattern cutting class using Clo3D in a CAD laboratory.

learning environment. Tutors also emphasise the environmental benefits of 3D digital design, highlighting its potential to reduce sampling, save material costs, and promote sustainable practices within the design process.

8.3.3 Progression and Integration in Subsequent Years

Following the mandatory first-year content, students in their second year have the option to continue working with Clo3D while receiving tuition in analogue methods. Additional in-person digital workshops are offered to facilitate further skills acquisition and application. Students are encouraged to integrate both analogue and digital techniques in their project work, fostering a hybrid approach to design.

In the final year, students gain greater autonomy, allowing them to independently develop their pattern cutting skills using their preferred methods. This progression equips graduates with a comprehensive toolkit to meet the evolving demands of the fashion industry.

8.3.4 Evaluating Student Perceptions

To assess the effectiveness of the teaching and learning methods, a mixed-methods survey was conducted. Students who opted to pursue digital pattern cutting in their second year were invited to participate. The survey included open- and closed-ended questions, with the latter rated on a four-point Likert scale. This approach provided both qualitative and quantitative insights.

Out of the 15 students who chose the digital pathway, eight responses were received. While the sample size was small, it consisted of students predisposed to digital practices, offering valuable insights into the experiences and attitudes of this group.

8.4 FINDINGS

The survey was analysed using descriptive statistics to present the quantitative results achieved for a series of statements related to their use of the digital pattern cutting application Clo3D, as seen in Figure 8.5.

This shows that the students in the sample are positive about the use of the technology in terms of understanding the digital pattern cutting method, its integration with analogue methods, and specifics such as the technical assessment of proportion and fit and its potential to enhance employability. Although most responses where positive, there were negative scores related to the efficiency of workflow and, interestingly, to associated sustainability credentials.

Following an analysis of the free text comments, this qualitative data gave rise to the following positive insights. Students acknowledged workflow efficiencies; being able to visualise their design on screen, assessing proportion, fit, and rectifying errors before committing to a real toile. One student reported that their confidence in pattern cutting had increased since using Clo3D, while another student stated that it was enjoyable to use. Further qualitative insights showed that students appreciated the employability context as the industry adopts 3D digital pattern cutting methods.

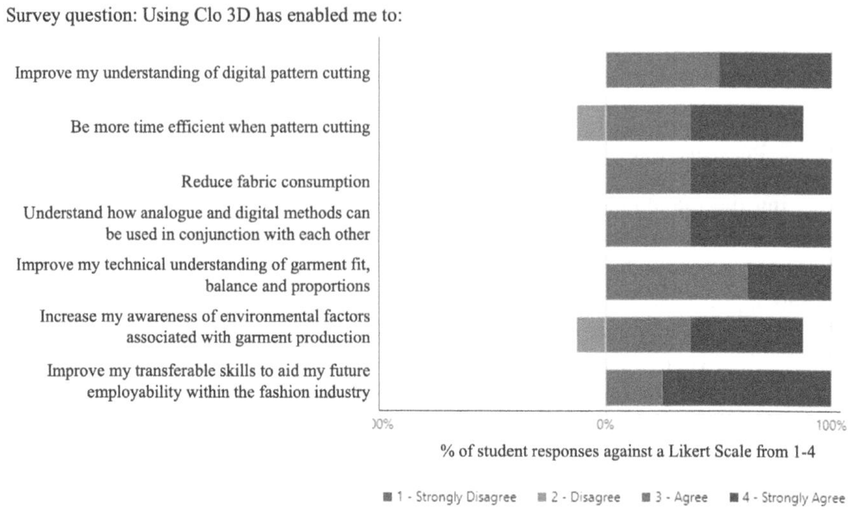

FIGURE 8.5 Quantitative data to describe students' opinions on a variety of issues related to their use of Clo3D for digital pattern cutting.

Most students found the software easy to learn; however, some students noted that the software takes time to learn, with many tools and techniques to remember.

When triangulated with the quantitative results, it showed that there was convergence with the qualitative insights for statements that referred to using the technology and future employability prospects. Both the qualitative and quantitative results showed that a small number of students found the software challenging to learn at the start of the block. It was interesting to see that there was no reference to sustainability credentials mentioned in the free text comments; however, the quantitative results showed that students were mostly in agreement that using 3D digital pattern cutting software had increased their awareness of environmental factors associated with garment production. This divergence provides an opportunity for a deeper investigation.

8.5 DISCUSSION

The findings reveal key themes related to the integration of digital pattern cutting into a traditionally analogue curriculum. Despite the limited sample, the study underscores the importance of aligning digital and analogue methods, as well as the potential of 3D tools to enhance efficiency and sustainability in the fashion design process.

8.5.1 Pedagogical Approach

Based on student feedback, the structured approach of introducing digital pattern cutting after a significant amount of traditional teaching has been well-received.

The digital teaching takes place in real time, in person, and follows the format of the traditional in-studio classes. This creates a conducive environment for effective and interconnected learning. It enables extensive scaffolding through peer-to-peer learning and VLE recordings and aligns with the pedagogic theories of constructivism, connectivism, the Zone of Proximal Development in the literature.

8.5.2 Developing Skills and Workflows

Undertaking analogue skills acquisition first enables students to appreciate the benefits of Clo3D, being able to view the silhouette, proportion, and style lines of a garment rapidly on screen from all angles. This, combined with their experience of manipulating physical fabrics, provides an understanding of how different fabrics drape and enables them to critically assess how realistic a Clo3D visualisation is. Student reflection on workflows revealed that this teaching method had increased their confidence in pattern cutting, enabling them to switch between different modes in their work. This can be seen in one of the sketchbook pages from alumna Sofia Garvao's final collection, as seen in Figure 8.6, where the garment is being assessed on the avatar for fit and proportion.

The tutors are fully supportive of using the software and are encouraging students to incorporate it into their daily practice. While proficient in other digital pattern cutting applications, the tutors have also undergone training to become familiar with Clo3D tools. This training has helped them understand the software's intricacies, allowing them to empathise with the students' challenges in learning this software.

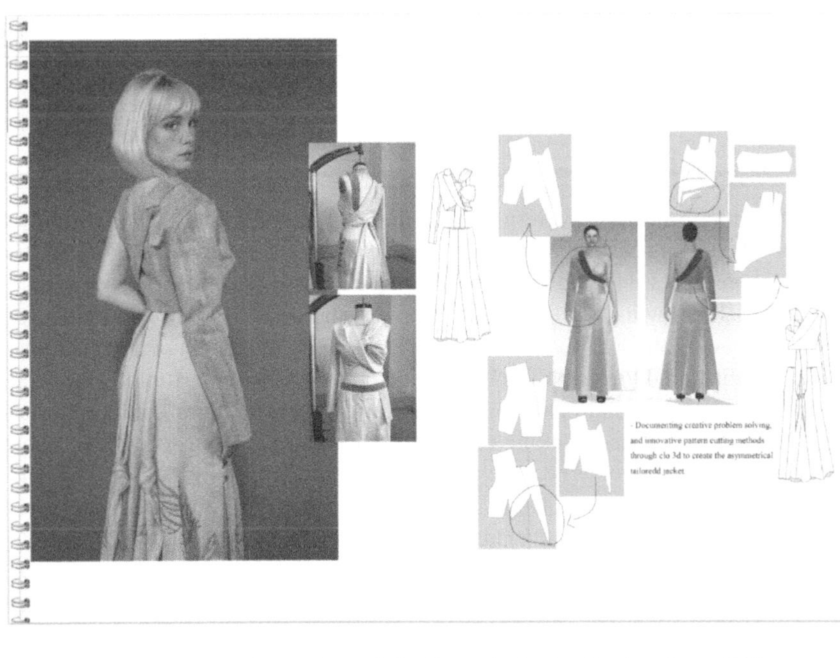

FIGURE 8.6 A sketchbook page from Sofia Garvao's final collection, June 2023.

However, even with this scaffolding in place, some students report that the software is challenging to learn at first. There is potential for further research in this area, specifically to explore how learning styles impact software skills acquisition. A comprehensive study of all students before and after the mandatory first year digital block would offer an opportunity to explore this further amongst a population of Gen Z learners.

8.5.3 Sustainability Credentials

Clo3D provides an excellent platform to highlight the iterative nature of the design process and the number of samples that can be developed before reaching a final design. Students appreciate the potential material savings in their processes; however, the evaluation showed there is an opportunity to consider how that translates into the commercial design process. This could be done by working with students to calculate fabric savings per design. Developing an understanding of carbon literacy, related carbon footprint of fabrics savings, and an exploration of energy consumption of a digital platform would provide evidence of the resulting environmental impact.

8.5.4 Employability

Students are aware of how in-demand these skills are, which is an extra employability incentive. There are placement and graduate opportunities for designers with Clo3D skills advertised. New careers are emerging that centre on the creation of garment visuals both for the design and buying processes. Although the sensorial aspect of the analogue prototype is currently not present in the digital representation, this is compensated for by the speed of production. As such, fast fashion brands are adopting this technology for this reason, which does raise the question about this technology's overall environmental impact.

8.5.5 Future Directions

The integration of digital pattern cutting into fashion education paves the way for transformative changes in pedagogy and professional practice. As digital tools become more accessible and sophisticated, future curricula may prioritise a digital-first approach, allowing students to experiment and iterate rapidly before engaging with analogue methods. Advancements in artificial intelligence, such as automated fit adjustments, could reshape workflows and provide new opportunities for creativity and innovation. Additionally, adaptive learning technologies could personalise the education experience, making complex software skills more accessible. Entirely virtual design careers may become more prominent, with roles focused on creating digital garments for the metaverse and AR platforms. This shift will likely influence educational programs to include modules that focus solely on developing virtual assets.

While 3D software offers significant advantages in speed, precision, and material efficiency, it lacks the tactile dimension central to the traditional process. In the analogue process, handling physical materials allows students to experience how fabric drapes, stretches, or resists, enabling them to refine their patterns through an iterative process of toile creation and adjustment. Digital tools simulate fabric properties with increasing sophistication, but even advanced software cannot yet replicate the sensory feedback of working with real textiles. This gap poses challenges for students in developing a nuanced understanding of how design decisions translate to physical garments.

Emerging research to develop interactive haptic experiences in immersive environments will also be a highly relevant advancement to digital pattern cutting, providing the designer with the ability to understand the handle of a fabric as well seeing it draping over a mannequin.

Educators must address this current limitation by emphasising the interplay between digital and analogue methods. Encouraging students to test their digital designs in fabric and critically compare outcomes ensures they develop both the technical and sensory skills essential for professional practice.

Finally, as digital workflows streamline production, critical discussions on the ethical and environmental implications of energy consumption will also become increasingly important. There will be a focus on sustainability metrics, such as tracking carbon footprints via material savings from using digital processes in addition to the energy used for data storage.

8.6 CONCLUSIONS

This study has shown that a structured integration of analogue and digital pattern cutting can provide an exciting dimension to fashion design education, providing an opportunity for a new method of working. The pedagogic approach adopted at De Montfort University demonstrated that delivering tuition for both analogue and digital pattern cutting through synchronous, face-to-face sessions offers strong scaffolding and was met with positive student feedback. Students appreciated the opportunities presented by developing workflows that integrate manual with digital methods. Students recognise that there are new career opportunities for graduates with these 3D digital pattern cutting skills and, to some extent, appreciate the potential for reducing the environmental impact of the design process. This study has opened further avenues for investigation, specifically around the challenges and issues connected with learning a digital pattern cutting software application and the wider consideration of this technology's sustainability credentials in the context of speed to market. With continued development, fashion education can prepare students for an evolving industry while encouraging sustainability and innovation.

REFERENCES

Aldrich, Winifred. 1996. *Fabric, Form, and Flat Pattern Cutting*. Oxford: Blackwell Science.

Almond, Kevin. 2016. The Status of Pattern Cutting. *The Journal of Design, Creative Process and the Fashion Industry*. https://doi.org/10.1080/17569370.2016.1147692.

Gonzalo, Ana, Holger Harries, Carlos Sanchez-Altable, and Coralie Villepeit. 2019. *Fashion's Digital Transformation: Now or Never*. McKinsey & Company. Available at: https://www.mckinsey.com/industries/retail/our-insights/fashions-digital-transformation-now-or-never.

Hamer, Robert, and Erik Van Rossum. 2010. *The Meaning of Learning and Knowing. Education Philosophy. Semantic Scholar* [PDF]. Available at: https://www.semanticscholar.org/paper/The-Meaning-of-Learning-and-Knowing-Hamer-Van-Rossum.

Selwyn, Neil. 2011. *Education and Technology: Key Issues and Debates*. 2nd ed. London: Bloomsbury.

Taber, Keith S. 2015. Meeting Educational Objectives in the Affective and Cognitive Domains: Personal and Social Constructivist Perspectives on Enjoyment, Motivation, and Learning Chemistry. In *Affective Dimensions in Chemistry Education*, 3–27. https://doi.org/10.1007/978-3-662-45085-7_1.

Taylor, Patrick. 1990. *Computers in the Fashion Industry*. Oxford: Heinemann Professional Publishing.

Wenger, Etienne. 2010. Communities of Practice and Social Learning Systems: The Career of a Concept. In *Social Learning Systems and Communities of Practice*, edited by Chris Blackmore, 1978. London: Springer. https://doi.org/10.1007/978-1-84996-133-2_11.

Vygotsky, Lev. 1978. *Mind in Society: Development of Higher Psychological Processes*. Cambridge, MA: Harvard University Press. https://doi.org/10.2307/j.ctvjf9vz4.

9 Digital Pattern Cutting and Simulation in Archival Garment Studies

Lucie Shilton

9.1 INTRODUCTION

This chapter examines an innovative approach that showcases archival garment research and explores the methodology employed in a recent study of digital garment simulation technology. Focusing on the reconstruction of a 1970s garment by fashion designer Bill Gibb, digital technologies CLO3D, Gerber AccuMark and Blender were combined to reconstruct the archival garment to reveal previously hidden pattern cutting methods. The presented methodology offers a structured approach to analysing and reconstructing archival garments within the evolving field of digital fashion and pattern cutting research. This process involved retracing the original garment's construction, recreating the garment's two-dimensional pattern and developing a three-dimensional digital simulation. Additionally, a method was devised to visually communicate the relationship between the garment's two-dimensional and three-dimensional forms, broadening pattern cutting knowledge for a wider audience.

Archival garment studies have been revolutionised by new developments in the field of digital archiving technologies. One significant benefit of this technology has been the preservation and conservation of valuable historical artefacts, providing free online access beyond the constraints of the object's physical location (Pecorari 2019). Through the visual deconstruction of the garment into its component parts, there is potential to supplement knowledge of existing museum collections and be a valuable educational resource. Such a process allows for a reinterpretation of the pattern cutting approaches used at the time of the garment's construction, revealing meaningful insights from the past. By bringing these insights to the consciousness of new audiences, there is the opportunity to enhance knowledge development in both fashion and archival industries. Despite the clear benefits of technological innovations in the field of digital pattern cutting and archival dress studies, there is little published research which outlines a methodological approach to this specialist area. Combining object-based analysis and practice-based research, this chapter shares a framework for digitising archival garments, to supplement existing museum collections and showcase the role of the pattern-cutter. By sharing the approach, this chapter presents a framework that can be adapted or used for other pattern-cutters in both commercial practice and for future research.

DOI: 10.1201/9781003594482-11

9.2 LITERATURE REVIEW

Access to fashion archives presents numerous challenges to designers, pattern-cutters and researchers in the study of fashion design and manufacture. Currently, only a small percentage of museum artefacts are available for display due to lack of physical space (McNulty 2019, 239), and in-person research appointments are restricted by limited availability and the geographical location of each museum (Kupresanin and Chen 2019). Further considerations for the fragility of older garments and the delicate nature of many archival pieces impose significant limitations on handling and display, further restricting opportunities for in-depth study, and constraining learning from these artefacts (The Museums and Galleries Commission 2001).

Historical fashion researchers Arnold (1966, 1977, 1985) and Kirke (1998) have made significant contributions to the field of archival garment studies by uncovering garment pattern cutting techniques through their analysis of notable historical artefacts. Their work serves as an invaluable resource preserving knowledge that underpins education and understanding in the field, revealing insights which might otherwise remain undiscovered or lost. Today's challenge is to build on this commitment to preserving pattern cutting knowledge by fully exploring the untapped potential of simulation technologies in this field.

Recent advances in digital archiving have transformed the preservation, conservation and dissemination of historical and contemporary artefacts, offering free online access that transcends the limitations of physical location (Pecorari 2019). Digitisation expands an object's reach, freeing it from the constraints of physical storage and enabling broader accessibility (Franceschini 2019). Initially limited to curators and in-person researchers, free online accessibility now overcomes geographical barriers and encourages the 'democratization of knowledge' (Pecorari 2019, 12), expanding opportunities for global engagement and knowledge exchange. International museums such as the Kyoto Costume Institute, the Metropolitan Museum of Art and the Victoria and Albert Museum have recently expanded the scope of their digital archives (McNulty 2019; Melchior 2019). By increasing the volume of their collections available online and providing high-resolution imagery, the museums have enhanced public accessibility and facilitated deeper engagement with their garment collections. However, despite this opening of access, McNulty (2019) highlights the limitations of 2D photography which constrain the viewer's perspective to fixed images. Research suggests that 3D solutions can enhance digital collections, as can be seen in Google's We Wear Culture platform and virtual exhibitions by the Brooklyn Museum which demonstrate the potential of 360° views, high-resolution imagery and interactive storytelling. This potential for technological innovations in the field offers exciting opportunities to the researcher and practitioner.

In the commercial fashion industry, three-dimensional garment simulation technology has become a valuable tool for creating virtual prototypes, enabling virtual fashion shows, digital fittings and improved communication, all of which contribute to the reduction of physical sampling (Arribas and Alfaro 2018). Recent advancements in digital pattern cutting technologies have significantly transformed industry practices, with studies (Arribas and Alfaro 2018; Wang and Lui 2020; Porterfield and Lamar 2021) highlighting the positive impact of these innovations. Driven by the challenges posed by COVID-19 (Baek et al. 2022), these technologies allow for the creation of 3D digital models that simulate physical garments on digital avatars, providing accurate representations for design exploration. This capability enables design decisions to be refined before physical sampling, enhancing communication between designers and factories and streamlining the product development process. Virtual prototyping encourages free experimentation and testing with digital materials, significantly reducing the need for physical samples, contributing to sustainable practice. Furthermore, the technology supports virtual fittings by using colour-coded 'fit maps' to identify and adjust areas of stress (Wang and Lui 2020).

An application of this technology can be implemented to create digital replicas, a technological reconstruction of an original garment, offering opportunities to experiment and manipulate archival garments (Martin and Ko 2011). Digital tools enable valuable sites of investigation with the potential for innovative approaches to garment design, preservation and sustainability. Moskvin, Kuzmichev and Moskvina characterise the development of digital replicas as a 'multidisciplinary area of research that combines cultural research, science-based methods of reconstruction and virtual reality' (2019, 1810). This complexity presents both challenges and exciting opportunities for contemporary pattern-cutters and designers, revealing insights during the reverse-engineering process.

Recent research explores inroads to the digitisation of historical garments. Moskvin et al. (2019) studied the digital recreation of an 1860s crinoline using CLO3D. Their research establishes a complex mathematical model to calculate the garment's hoop diameters, as previous studies had developed solutions feasible for simpler skirt construction but did not account for the challenges of a supportive frame. While applicable to similar historical hooped skirts, the scope is limited to addressing the unique requirements of this garment. Furthermore, the complexity of the calculations poses a challenge, making the method less accessible for broader application or adaptation to other garment types.

A similar study utilises CLO3D technology to replicate the pattern, colour and texture of a Qing Dynasty dragon robe (Wang et al. 2021). Data collected from appropriate literature determined the garment's size and aided the development of a 2D digital pattern. Due to the nature of this garment, automatic arrangement points could not be utilised, making the garment more challenging to simulate. However, experimentation with the system functions resulted in a successful simulation by altering the avatar's arm position and temporarily dividing the garment across the shoulder line. This underscored the advantages of using a rigged avatar, which allowed for joint manipulation to facilitate garment placement while avoiding collision issues. The outcome demonstrated 'a high degree of consistency' (2021, 5) when compared to the original museum garment, indicating that digitisation can serve as a reliable method for preserving and recreating culturally significant artefacts.

A further paper experimented with CLO3D to outline the virtual simulation process of 12 Yue opera costumes, enabling audiences to 'observe traditional costumes more easily and in greater detail' (Liu et al. 2022, 2). These culturally significant garments, often photographed flat, now could offer the possibility of being viewed in 360°. However, the simplistic X-shaped garments used in this study are relatively straightforward to replicate, highlighting the need for further research to apply this approach to garments with more complexity. While these examples provide valuable insights into this emerging field, their approaches are tailored to address specific historical garments' silhouettes, limiting their applicability to a wider range of garments. Further research is required to test and refine methods for further use. The studies also highlight the need for a clearly defined model for garment digitisation that can be applied more broadly, a gap that this study seeks to address.

9.3 METHODOLOGY

The research adopted a practice-based methodology which included a qualitative, interpretivist approach combining archival study with studio enquiry. Commonly employed in creative art and design fields, practice-based research generates new knowledge through practice, positioning hands-on activity at the centre of the work (Vear 2021, 29). Gauntlett (2022) emphasises making and experimenting as 'necessary and essential parts of the research process'. A systematic approach was conducted to collect and analyse data. The operational flowchart (see Figure 9.1) outlines the step-by-step research process which followed two key stages: Stage 1 – Analysis, and Stage 2 – Interpretation. Through a series of observations, images, measurements, technical drawings and field notes, a rich source of qualitative data was obtained. This data informed the experimentation and development of a physical and digital replica and the creation of a pattern cutting animation. Lastly, at the final part of the process, the physical and digital replicas were evaluated to the original garment and analysed through an evaluation framework.

In this research, the principal object of study was a 1970s Bill Gibb garment, housed in the fashion archive at the Central Saint Martins Museum and Study Collection. The artefact was analysed using object-based analysis and interpreted using a series of manual and digital experimental methods specified later. Multiple technologies integral to this research process were used as the foundation of the methodology which included the two-dimensional pattern cutting functions in Gerber AccuMark, the virtual garment simulation capabilities in CLO3D and the 3D modelling and animation possibilities in Blender. This section outlines the systematic approach applied to collect, analyse and interpret data from the original artefact to develop a digital pattern cutting replication.

9.3.1 Stage 1 – Analysis

Object-based analysis was conducted, following specified handling protocols to gather information on the material quality, construction techniques and garment finishes. A framework for capturing essential information was adapted from Mida and Kim's 'Checklist for Observation' (Mida and Kim 2015, 216–219), noting important details on the garment's style, materials, labels and construction techniques. Supported by a series of close-up photographs (see Figure 9.2) and written observations, an understanding of the garment's pattern cutting approach was reinforced. This format was essential in facilitating effective data collection.

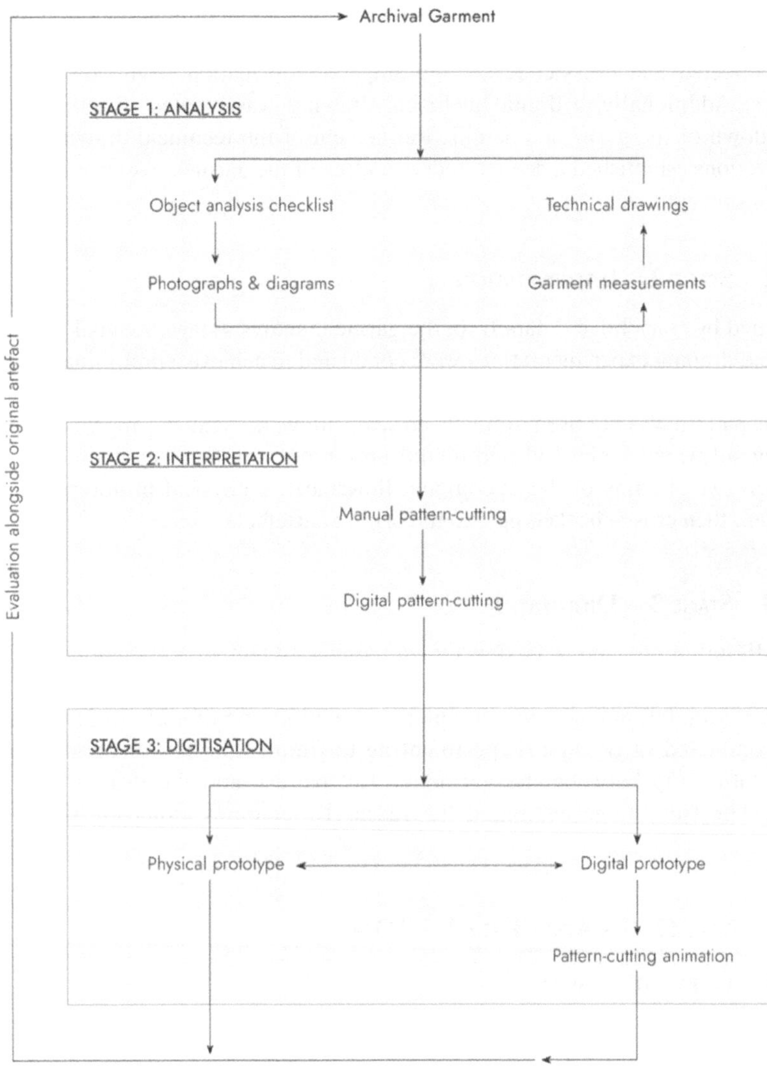

FIGURE 9.1 Operational flowchart.

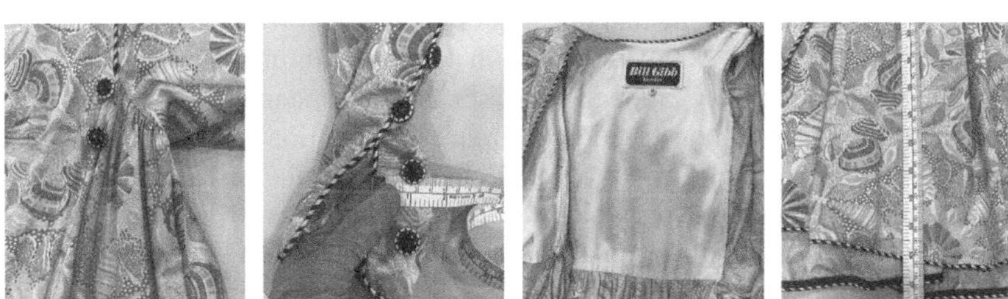

FIGURE 9.2 Close-up photographs from in-person handling session.

Through contact with the original artefact, a sequence of detailed measurements was conducted and entered to a measurement chart. Combined data from the archive was collected and cross-checked, building on information from previous observations. Additionally, a digital technical drawing was created, forming a visual breakdown of the garment's details. Refinement of the technical drawing between observations established a deeper understanding of the garment which aided the pattern cutting approach.

9.3.2 Stage 2 – Interpretation

Informed by consolidated data from the garment analysis stage, manual pattern cutting and draping experimentations were conducted to achieve a preliminary garment pattern.

The pattern was digitised, refined and seam allowance added using Gerber pattern cutting software. A physical and digital prototype were conducted simultaneously through two streams of the operational flowchart. A physical prototype was constructed, then cross-checked next to the original artefact.

9.3.3 Stage 3 – Digitisation

The digital prototype was developed using CLO3D and the model adjusted accordingly, informed by learning from the physical prototype. Once the static garment simulation stage was complete, an exploration of a 2D-to-3D animation was conducted in Blender, experimenting to find a solution for communicating the relationship between the garment's flat pattern and the three-dimensional form. The final stage evaluated the research outcomes alongside the original artefact.

9.4 OUTCOMES AND DISCUSSION

9.4.1 Garment Analysis

The initial garment analysis phase involved a tactile discovery process, gradually building knowledge of the artefact in stages. Taking time to observe the garment as a whole was essential for developing a grounded understanding before systematically dissecting it into its component parts. The two initial in-person research sessions permitted gentle turning and opening of the garment while laid flat, providing opportunities for deeper exploration. This hands-on examination provided further opportunities for discovery, allowing for a deeper understanding of the garment's pattern cutting approach, construction techniques, fabrication and finishes. However, the study of this archival dress presented challenges to the research process. For example, the handling protocols which were imposed in order to protect the garment resulted in limited manipulation. As is common practice in an archival setting, the garments can often only be handled using a gloved hand, with turning and opening guided by the archivist. These factors restricted the possibility to gain a full understanding of the garment's textile quality. As a result, understanding the artefact's material qualities was limited to visual observation and the researcher's prior knowledge and experience with similar textiles.

The technical drawings generated from collected data provided a deeper understanding of the garment and emphasised the relationships between its individual pattern pieces. This step aided processing of technical information prompting further questioning. This cycle allowed for clarification in the second in-person handling session forming a link stage between the original artefact and the pattern cutting development.

Measurements were systematically collected by working through the garment from top to bottom, then front to back. This structured approach allowed for the fluid extraction of measurements specific to the garment's unique details. Certain aspects posed challenges, such as determining the sleeve head width due to the gathered fullness at the sleeve crown. In such cases, measurements were taken partway down and recorded as estimates in a collated chart. The chart facilitated vital cross-checking and refinement, underscoring the importance of multiple observations to ensure data accuracy, which was an essential foundation for precise pattern cutting development.

9.4.2 Pattern Cutting Development

The development of the garment's pattern was guided by the garment analysis data and key measurements, which were mapped onto a mannequin using draping tape. This manual draping approach proved valuable in visualising the evolving three-dimensional form, refining the style lines and bridging the transition between the original artefact and its digital reconstruction. However, challenges arose in correlating measurements to the 3D form, such as determining the depth of the crossover point. In these instances, the most reliable measurements were prioritised, and previous pattern cutting knowledge was applied to inform decision making.

Traditional methods for reproducing garment patterns typically involve pinning and tracing the shape onto a new flat material. However, due to archival restrictions, an alternative method of measuring and replotting the pattern was employed. While this reverse-engineering method minimised interaction with the artefact, it is inherently less precise than direct pattern tracing. Nonetheless, it was the most appropriate technique within an archival context. Further, limitations to accuracy stemmed from restricted access to the garment. In this study, the researcher was granted three observation sessions, limiting opportunities for cross-checking when questions arose. Future research would benefit from in-house access to the archive, allowing for a more iterative and precise approach to pattern reconstruction.

The body and sleeve shape were digitised into Gerber technology, allowing for further refinement and development through digital pattern cutting tools. Automated functions streamlined the process, such as the use of the slash-and-spread tool to add fullness to the skirt and sleeve, to reflect the original garment's measurements.

Key benefits of digital pattern cutting in this phase included precise cross-checking of measurements, quick pattern adaptations and automatic application of seam allowance. Additionally, digital marker-making software enabled quick material requirement assessments, providing valuable insights into the original garment's fabric consumption and fabric wastage percentages. This secondary outcome proved both insightful and highly relevant to contemporary discussions on sustainability in fashion.

Physical prototyping allowed for testing of the replicated garment pattern and construction methods. Analysing the prototype on a live model provided valuable insights into the garment functionality in a real-life context, for example the drape of the garment. Additionally, discussions with the lead archivist led to further experimentation, comparing both the artefact and prototype side by side on mannequins. Observing the artefact in its three-dimensional form proved invaluable to the research and analysis process. This parallel assessment enabled further refinements to the pattern cutting, reinforcing the physical prototype as a crucial component of the research. In this instance, development of the physical garment prototype served as a vital bridge between the digital pattern and the physicality of the artefact, underscoring the importance of iterative engagement between the physical and digital realms during the research process.

9.4.3 Digital Simulation

During the garment simulation phase, pattern pieces were arranged around the digital avatar in CLO3D, with digital stitching applied to establish an initial silhouette. However, it became evident that not all sewing processes could follow the same construction order as the physical prototype. For instance, seam and hem allowances were unnecessary in the digital replication and were subsequently removed. Similarly, the top stitching – which is typically applied during the physical construction process – was applied decoratively after the main digital reconstruction was completed and served no structural implication. This phase furthermore highlighted the importance of digital proficiency in effectively manipulating standard technical functions to achieve accurate representation. For example, in some cases, garment patterns required adjustment or exclusions to optimise visualisation; this was found when layering pieces in CLO3D. A key challenge arose in the application of digital piping, as the fullness of the pattern pieces caused overlapping material, which was problematic for the software to follow. At the time of writing, limitations in the software prevented a seamless application, requiring an alternative solution. Initially tested in the 2D window, the issue was resolved by carefully following the draped garment's contours in the 3D window. Encounters with software limitations are likely to present ongoing challenges in complex garment simulations. As a result, experimentation and the development of alternative solutions remain a crucial aspect of the process until the software evolves to meet the necessary technical specifications.

The buttons on the original garment were a distinctive design element – a recurring signature of Gibb's aesthetic – but were not available in the standard CLO3D component library. To achieve accurate representation, custom buttons were modelled to mirror the original artefact (see Figure 9.3). This process was carried out using Blender, where the three-dimensional shape was re-created by modelling and duplicating segments to reflect the symmetrical design. UV mapping was then used to flatten the object's faces into a 2D image, allowing colour and pattern to be applied in Adobe Photoshop. While the button development was complex and time-consuming, it proved invaluable in achieving precise finishing details, enhancing the authenticity of the digital reconstruction.

Once the digital garment was completed, a rendering process allowed for the extraction of alternative viewpoints that would not be easily possible when displaying the original artefact. For example, disabling the avatar-view provided new and insightful perspectives, enabling exploration of the garment from the ground-up or bird's-eye orientation (see Figure 9.4). This unexpected outcome suggests potential applications for this research. Further experimentation included testing a transparent render by substituting the original digital textile with organza and reducing opacity. This exploration with alterative materials highlights the advantages of digital technology as a research tool, revealing hidden layers of the garment's internal structure.

When comparing the physical and digital construction processes, many digital techniques proved more efficient, such as digital stitching and the instant mirroring of construction from one side of the garment to the other. However, certain digital

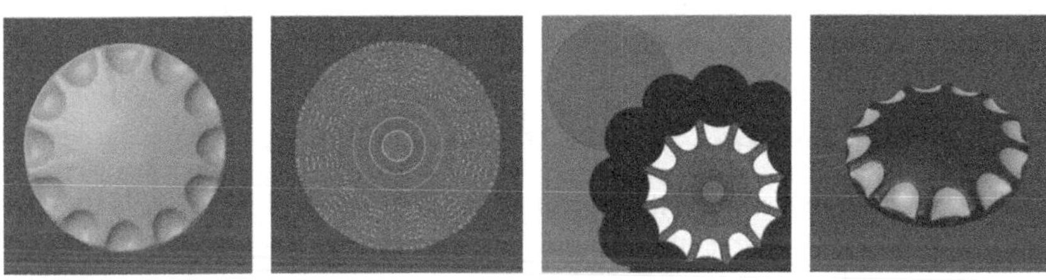

FIGURE 9.3 Custom button design developed using Blender.

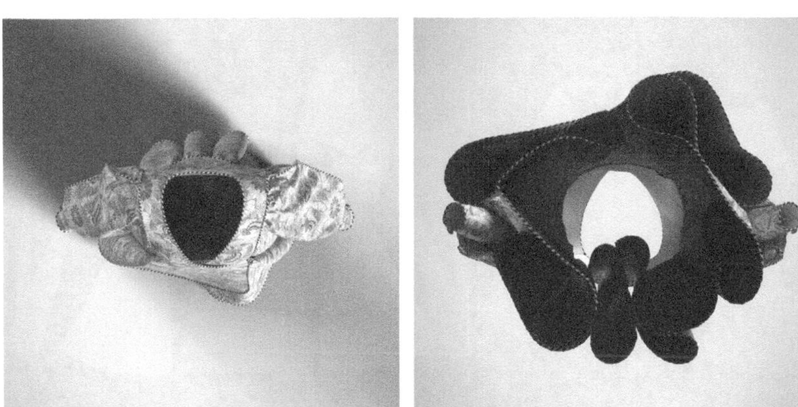

FIGURE 9.4 Digitally rendered garment viewed from a bird's-eye view and ground-up perspective.

processes were more complex and time-consuming than the physical prototyping. For example, due to the width of the flared skirt panels, significant time was required to manage the fluid drape of the skirt while avoiding collision issues. Additionally, as previously mentioned, modelling the digital buttons was particularly time intensive. In such cases, careful prioritisation is necessary to ensure the most effective use of time and resources, balancing accuracy within the constraints of budget and project timelines.

9.4.4 Animation Development

Through experimentation with Blender software, a workflow was developed to integrate the 2D and 3D digital pattern using 'shape key' functions. The process began with the 3D version of the file, from which an additional version was exported with pattern pieces reset to their 2D arrangement. Extensive trial and error testing revealed that maintaining the file's structural integrity was crucial for the software to successfully merge both objects. Therefore, all garment pattern baselines were locked to avoid any unintentional changes to the garment's mesh when repositioning pattern pieces from 2D to 3D. Initial testing showed that morphing between the 2D and 3D formats did not produce a natural transformation in the mid-frames. To address this, an additional mid-way simulation was introduced, allowing greater control over the animation and resulting in a smoother, more seamless transition (see Figure 9.5).

9.4.5 Project Evaluation

The evaluation phase assessed the digital replica and physical prototype in comparison to the original artifact, considering both limitations and potential improvements. Feedback from discussions with museum archivist was positive, highlighting the effectiveness of pattern animations in enhancing understanding of pattern cutting

FIGURE 9.5 Development of 2D-to-3D animation.

techniques and illustrating how specific pattern shapes interact and conform to the human form. A key point of discussion emphasised the importance of honouring the authenticity of the original garment while exploring the potential for applying this approach in museum settings. The findings underscored the value of digital animations as a tool for enrichment and storytelling, offering new ways to engage audiences with archival fashion.

9.5 CONCLUSION

This research examines emerging applications of digital fashion technologies, offering new avenues for archival practice and the study of dress. It is essential to recognise that digital replicas serve as complementary tools rather than a direct substitute for physical artefact observation. While digital reconstruction cannot fully replace the experiential knowledge gained from in-person interaction, it provides valuable opportunities to supplement and enhance museum collections.

In this research, direct handling was crucial for accurately extracting measurements and developing a comprehensive understanding of pattern cutting techniques. However, access was limited to three viewings, reducing opportunities for frequent cross-referencing in a studio setting. Working in-house with unrestricted access to archival resources would facilitate deeper interaction and improve accuracy. The physical prototype played a vital role in bridging the gap between the digital pattern and the original artifact, highlighting the importance of iterative engagement between physical and digital forms. Moreover, the time-intensive nature of garment observation and pattern replication underscored the value of maintaining digital records, ensuring accessibility for future researchers.

This research connects archival study, two-dimensional pattern cutting and three-dimensional garment simulation, advancing scholarship in this field. The proposed framework for pattern cutting digitisation presents exciting opportunities for archival practice by revealing pattern cutting and construction details hidden within historical garments, which broadens access on a global scale. Notably, this methodology offers a means to recreate fragile or damaged artefacts unsuitable for regular handling, providing an effective approach for fashion digitisation and archival study that can be refined and replicated in future professional practice. By unlocking pattern cutting knowledge embedded within the garment's construction, this research facilitated wider dissemination of past pattern cutting techniques in an accessible format. The potential for collaboration within a global research network would enable the examination and digitisation of pattern cutting practices from worldwide locations. The ongoing exploration of pattern cutting communication reinforces the significance of this craft, fostering international knowledge exchange across the fashion, costume and archival industries.

Supplementary material:

Animation links:
https://youtu.be/_pusM3UB-T0
https://youtube.com/shorts/6mmrWG2SBW8

REFERENCES

Arnold, Janet. 1966. *Patterns of Fashion 1: Englishwoman's Dresses and Their Construction c. 1660–1860*. Reprint. London: MacMillan, 1977.

Arnold, Janet. 1977. *Patterns of Fashion. 2: Englishwomen's Dresses and Their Construction 1860–1940*. Reprint. London: MacMillan.

Arnold, Janet. 1985. *Patterns of Fashion: The Cut and Construction of Clothes for Men and Women C1560–1620*. London: New York: Macmillan.

Arribas, Veronica, and José A. Alfaro. 2018. '3D Technology in Fashion: From Concept to Consumer'. *Journal of Fashion Marketing and Management: An International Journal* 22 (2): 240–251. https://doi.org/10.1108/JFMM-10-2017-0114.

Baek, Eunsoo, Shelley Haines, Omar H. Fares, Zhihong Huang, Yuwei Hong, and Seung Hwan Mark Lee. 2022 December. 'Defining Digital Fashion: Reshaping the Field via a Systematic Review'. *Computers in Human Behavior* 137: 107407. https://doi.org/10.1016/j.chb.2022.107407.

Franceschini, Marta. 2019. 'Navigating Fashion: On the Role of Digital Fashion Archives in the Preservation, Classification and Dissemination of Fashion Heritage'. *Critical Studies in Fashion & Beauty* 10 (1): 69–90. https://doi.org/10.1386/csfb.10.1.69_1.

Gauntlett, David. 2022 January 12. *Practice-Based Research: A Simple Explainer*. https://davidgauntlett.com/research-practice/practice-based-research-a-simple-explainer/.

Kirke, Betty. 1998. *Madeleine Vionnet*. San Francisco, CA: Chronicle Books.

Kupresanin, Andrew, and Belinda Chen. 2019 December 10. 'Fashion Beyond Physicality'. *Medium* (Blog). https://medium.com/@superbureau/fashion-beyond-physicality-486d71b05dae.

Liu, Kaixuan, Shunmuzi Zhou, Chun Zhu, and Zhao Lü. 2022. 'Virtual Simulation of Yue Opera Costumes and Fashion Design Based on Yue Opera Elements'. *Fashion and Textiles* 9 (1): 31. https://doi.org/10.1186/s40691-022-00300-0.

Martin, Kathi, and Hyeong-Seok Ko. 2011. 'Imagining Historic Fashion: Digital Tools for the Examination of Historic Dress'. In *2011 Second International Conference on Culture and Computing*, 51–56. https://doi.org/10.1109/Culture-Computing.2011.18.

McNulty, Roz. 2019. 'Future Applications of Digital Clothing for Historical Costume: The Past, Present and Future of Fashion'. In *Proceedings of EVA London 2019*. https://doi.org/10.14236/ewic/EVA2019.45.

Melchior, Marie Riegels. 2019. 'Digital Fashion Heritage: Understanding Europeanafashion.Eu and the Google Cultural Institute's We Wear Culture'. *Critical Studies in Fashion & Beauty* 10 (1): 49–68. https://doi.org/10.1386/csfb.10.1.49_1.

Mida, Ingrid, and Alexandra Kim. 2015. *The Dress Detective: A Practical Guide to Object-Based Research in Fashion*. London; New York: Society of Antiquaries of London, Bloomsbury Academic, An Imprint of Bloomsbury Publishing Plc.

Moskvin, Aleksei, Victor Kuzmichev, and Mariia Moskvina. 2019. 'Digital Replicas of Historical Skirts'. *The Journal of The Textile Institute* 110 (12): 1810–1826. https://doi.org/10.1080/00405000.2019.1621042.

The Museums and Galleries Commission. 2001. 'An Illustrated Guide to the Care of Costume and Textile Collections'. *Journal of the American Institute for Conservation* 40 (3): 267. https://doi.org/10.2307/3179883.

Pecorari, Marco. 2019. 'Fashion Archives, Museums and Collections in the Age of the Digital'. *Critical Studies in Fashion & Beauty* 10 (1): 3–29.

Porterfield, Anne, and Traci A. M. Lamar. 2021. 'A Framework for Incorporating Virtual Fitting into the Costume Design and Production Process'. *International Journal of Fashion Design, Technology and Education* 14 (1): 91–100.

Vear, Craig. 2021. *The Routledge International Handbook of Practice-Based Research*. Milton, United Kingdom: Taylor & Francis Group. http://ebookcentral.proquest.com/lib/ual/detail.action?docID=6818293.

Wang, Xu, Yue Li, and Xiaogang Wang. 2021. 'A Study of Replication of Qing Dynasty Dragon Robes Based on CLO3D Technology'. *Journal of Physics: Conference Series* 1986 (1): 012069. https://doi.org/10.1088/1742-6596/1986/1/012069.

Wang, Y, and Z. Lui. 2020. 'Virtual Clothing Display Platform Based on CLO3D and Evaluation of Fit'. *Journal of Fiber Bioengineering and Informatics* 13 (1): 37–49. https://doi.org/10.3993/jfbim00338.

Part C

Engineered Reality in Fashion

10 Virtual Reality, Traditional Textiles and Contemporary Consumers

Karen Cross, Lyndsay Mesjar, Rong Zheng, Yang Jiang, and Josie Steed

10.1 INTRODUCTION

This chapter presents research undertaken for Augmented Fashion (AF), an interdisciplinary UK-China research project funded by the Arts and Humanities Research Council involving academics and industry partners from the fashion, textiles and computing science disciplines in both the UK and China. The project sought to explore ways to communicate the craftsmanship, heritage, value and sustainability of traditional fashion and textile products, exploring the use of immersive technologies to engage contemporary audiences with these traditional products and encourage them to consume fewer fast, disposable fashion items. Initial research established that fashion brands were using virtual reality (VR) or augmented reality (AR) in three ways: to show products via 3D modelling and virtual fashion shows, to enable virtual try-on and support purchase decision-making (e.g. using customisable avatars, AR tracking or body scanning) and to provide an experience, enabling customers to interact with the brand socially, without specific purchase intent. There were few examples of fashion brands using immersive technologies to convey craftsmanship, heritage or provenance (Cross et al. 2021). Thus, the AF project focused on developing human-computer interactions that communicated culture, place, provenance and the importance of the human hand in traditional fashion and textile production, aligning with the United Nations Sustainable Development Goal 12: responsible consumption and production (United Nations 2015).

Harris Tweed® is a brand that epitomises craftsmanship, heritage and provenance (Serdari 2018). It has been woven on non-power treadle looms in weavers' homes on the Scottish Hebridean island of Lewis and Harris for centuries (Platman 2011), and the Harris Tweed Orb is the oldest British trademark in continuous use (Wilson et al. 2021). Harris Tweed Hebrides (HTH) is the largest producer of authentic Harris Tweed® and was a key industry partner on the AF project. HTH seeks to balance the legacy of their craft and the sustainability of their 100% wool, small-batch production, with innovation and exposure to new markets to ensure sustained employment for their remote, island-based workforce (Steed et al. 2023), thus aligning with the United Nations Sustainable Development Goal 8: decent work and economic growth (United Nations 2015).

DOI: 10.1201/9781003594482-13

Prior to the project, HTH utilised traditional marketing methods to reach both business-to-business and end-user consumers, including face-to-face interactions at trade fairs and showrooms, printed brochures and leaflets, their website and social media (mainly Instagram, Facebook and LinkedIn). Keller (1999) notes the challenge traditional brands such as Harris Tweed® face in gaining relevance with new customer cohorts, especially where the brand already has strong user image associations. There is broad agreement in recent literature that effective omni-channel marketing is key as consumers increasingly make fashion and clothing purchases across multiple channels (Cross et al. 2023) and that there is increasing interest in VR as a marketing and promotional tool (Hollebeek et al. 2020; Kim et al. 2022). As an industry partner on the AF project, immersive interactions were developed as a potential new marketing medium for the HTH brand, and viewer reactions to the content were tested.

10.2 METHODOLOGY

This research utilised an interpretive research paradigm, acknowledging the possibility of multiple meanings when interpreting fashion (Barnard 2014). Adopting a qualitative methodology, the research was conducted via a survey that elicited response to two pieces of content, a traditional (non-immersive) brand video and a semi-immersive branded 3D virtual world. The survey questions were designed using brand identity theories, specifically Kapferer's (2012) Brand Identity Prism and Keller's (2019) Consumer-Based Brand Equity (CBBE) Pyramid, which are both widely used within brand research. Keller's CBBE provides a model for businesses to build brand equity from a starting level of brand *Salience*, or identity, through to peak level of brand *Resonance*, where customers are in a positive relationship with the brand, and advocate for it. Kapferer's Brand identity Prism identifies six key aspects of brand identity, including the 'picture of sender' (encompassing the *Physique* and *Personality* that the brand communicates) and the 'picture of recipient' (how the brand portrays its target audience and how the consumer sees themself in the brand). Finally, as sustainability was a key focus of the AF project, some of Kim and Oh's (2020) sustainable key word associations for fashion brands were utilised as pre-defined choices within the survey.

The research did not seek to gauge user experience in terms of the media or platform; rather, it focused on the attitudes towards the brand engendered by the content. For the traditional content, the Harris Tweed Hebrides (HTH) Fabric of Fashion video was used. This one-minute video combines real-world imagery, narration and music to showcase the stunning landscapes of the Outer Hebrides that inspire the colours and textures of the woven Harris Tweed® cloth. Clips of the cloth production and authentication of the finished cloth by the Harris Tweed Authority are included. For the immersive content, a 3D virtual world combining cultural elements of Shanghai and Scotland was created using Unity software by artist Dr John Walter, viewable via VR headsets as a pre-recorded fly-through. This cross-cultural amalgamation plays with scale, mixing elements of the real with more fanciful and fun imagery, with messages about sustainability and the production of Harris Tweed® embedded (see Figure 10.1). The immersive experience was limited to a pre-recorded fly-through of the virtual world, ensuring all respondents viewed the same content in the same way.

FIGURE 10.1 Screenshots of traditional video content and immersive content.

Based on Slater and Wilbur's (1997) framework for immersive virtual environments, the *level* of immersion in John Walter's 3D virtual world can be described as high, as there were limited signals indicating the presence of devices in the physical world (in this case, only the wearing and weight of the VR headset). The *surrounding* aspect of immersion can also therefore be described as high, engendered by wearing the headset and only seeing and hearing the virtual content (excluding the real world). The *vivid* aspect of immersion was moderate, as the virtual world was magical, rather than realistic (Sherman and Craig 2018), in that it employed cartoon-like, stylized representations. An issue with utilising a pre-recorded fly-through was that some respondents found the experience too fast, negatively impacting physical comfort (IxDF 2023) in the form of cyber sickness (Martirosov et al. 2022); this moderate *matching* aspect of the immersion (Slater and Wilbur 1997) meant the visual experience did not match the proprioceptive feedback (respondents could look around within the headset by moving their head but remained on the fly-through journey, unable to control speed of travel or direction and with no use of hand controllers required). Thus, in terms of the 'ultimate experience of telepresence' (Boyd and Koles 2019, 442), the respondents' immersive interactions were very limited, categorising this as a semi-immersive, low-involvement (Kim et al. 2022) experience.

Respondents were asked to watch the traditional video and asked a series of questions, to gauge their perceptions about the brand's image, its core principles (e.g., quality, sustainability, fashionability) and who the brand was aimed at (target customer). They were then asked to view the immersive content and answered the same questions again, allowing direct comparison of the responses to the traditional versus the immersive content. There were two periods of data gathering: one in Scotland and one in Shanghai. A full analysis of the Scotland-based data can be found in Cross et al. (2023). Jisc Online Surveys was used to build and administer the survey in the UK, chosen as a university-approved, GDPR-compliant platform. The survey was translated into Chinese and implemented via WeChat in China. Key words from the respondents' qualitative responses were uploaded to NVivo qualitative data analysis software to identify key themes. Minimal data cleaning took place (e.g. landscape and landscapes were cleaned to landscape, luxury and luxurious to luxury). The themes are presented as a series of comparative word clouds.

10.2.1 Participant Overview

Fifty-two responses were gathered from Chinese fashion consumers (university students, Mode Shanghai Fashion week textile trade fair attendees and visitors to the Shanghai Textile Museum). Thirty-eight responses were gathered from UK participants: a mix of computing students, fashion management students (university and college level), fashion and textile students and attendees at a CreativeTech Scotland gathering. Most respondents can be described as contemporary consumers within the Generation Z generational cohort, born between 1997 and 2012 (Francis and Hoefel 2018). Most respondents (92% in both countries) described themselves as 'interested' or 'very interested' in fashion. This was expected as they had been sampled purposively (Flynn and Foster 2009) to be a key informant sample (Marshall 1996), as interest in fashion and style was relevant to the research. Eighty-seven per cent of the Shanghai-based and 71% of the Scotland-based respondents were female. In terms of attitude towards fashion and clothing, 53% of the Scotland-based and 46% of the Shanghai-based respondents indicated they like to be innovative with what they wear; 35% of the Shanghai-based and 24% of the Scotland-based respondents like to update their wardrobes with fashion trends. A much smaller percentage (18% of the Scotland-based and 12% of the Shanghai-based) indicated a more utilitarian attitude towards clothing purchase by only shopping when they needed a specific item of clothing. Thus, most respondents align with the Innovator, Early Adopter and Early Majority categories of Rogers' (1983) Diffusion of Innovation model, demonstrating that the respondents were more fashion forward than average consumer samples. Whilst user experience with immersive technologies was not a key focus of the survey, an unexpected finding of the study was the limited experience of using VR headsets among the respondents, with 29% of the respondents based in Shanghai and 29% of the respondents based in Scotland indicating they had never used a VR headset, and several respondents showing unwillingness to do so.

10.3 RESULTS AND DISCUSSION

10.3.1 Comparing Brand Salience Elicitations

Firstly, respondents were asked what their general impressions of the brand were (see Table 10.1). Overall, there was a good level of *Salience* (the first level of Keller's (2019) CBBE model) evident from the responses to the traditional video, with words such as 'heritage', 'classic' and 'traditional' prominent. It is interesting to note that the Shanghai-based respondents lacked awareness of the Scottishness of Harris Tweed®, perhaps reflecting on a lack of general awareness of Scotland as a place.

A substantial difference can be seen in the responses to the immersive content (Table 10.2), with the brand now seen as 'modern', 'interesting' and 'futuristic', but with lowered sense of prestige or luxury evident in the Scotland-based responses

TABLE 10.1
What Are Your General Impressions of the Brand Being Promoted in This Content?

Comments on Traditional Content (Shanghai)

Comments on Traditional Content (Scotland)

TABLE 10.2
What Are Your General Impressions of the Brand Being Promoted in This Content?

Comments on Immersive Content (Shanghai)

Comments on Immersive Content (Scotland)

(the word 'cheap' emerges). It is also interesting to note the heightened awareness of sustainability evident in the Shanghai-based responses, through words such as 'sustainable', 'natural' and 'handcrafted'. This may be due to the Chinese government's substantial focus on sustainability in recent years (Li et al. 2024).

Qualitative comments indicated appreciation for the sense of nature, landscape and heritage conveyed in the traditional video. There was, however, some negativity towards the immersive content, with respondents finding it, 'too busy and unfocused', 'confusing and distracting'. One respondent stated, 'nothing in this video seems professional, something you'd come to expect from a brand with such a rich history'; another stated, 'I absolutely hated [the immersive content]. . . . I have no idea what brand is being represented'. A more measured response noted that they did not 'think it aligns with their brand identity', being too futuristic. Positive responses towards the immersive content noted that it 'showed the brand's story in a creative and interesting way', that it was a 'very new experience' with 'strong visual effects' and that it 'gave a new perspective of the brand', making it feel more modern. Overall, there was more negativity towards the immersive content from the Scotland-based respondents; this may be due to cultural communication differences or differing levels of technology acceptance.

10.3.2 Comparing Core Principles Associations

Respondents were asked what they felt the brand's core principles were. This relates to the *Culture* element of Kapferer's (2012) Brand Identity Prism, which encompasses a brand's values and principles. The traditional content elicited core principles of 'tradition', 'craftsmanship', 'heritage' and 'quality', with the modifier 'high' often being used along with 'quality' (see Table 10.3). As seen in the previous question, the Shanghai-based respondents were slightly more focused on quality, with several mentioning inheritance of skills and craft. Locality or location was a key theme within the Scotland-based

TABLE 10.3
What Do You Feel Are the Core Principles of the Brand?

Comments on Traditional Content (Shanghai)

Comments on Traditional Content (Scotland)

responses, with some highlighting the need to support 'local businesses', an appreciation that everything about Harris Tweed® was kept within Scotland and stressing the 'importance of where it is made'. These findings differ from the Shanghai-based responses, perhaps again because they lacked familiarity with Scotland as a place.

'Tradition' and 'craftsmanship' also emerged in the responses gathered after the immersive content was watched, along with new themes of 'futuristic' or 'sci-fi' (abbreviation of science fiction), and 'fashionable' (see Table 10.4). One Scotland-based respondent felt the brand's principles were 'taking it into the future', one described it as 'future-facing', and another noted the immersive content 'echoes the same principles' as the traditional content, but 'with a slight twist towards a more modern way of working'. However, some respondents noted the immersive content felt 'slightly forced' and at odds with the Harris Tweed® brand. Overall, there was less certainty from the Scotland-based respondents, with several stating that they were 'not sure', 'confused' or had 'absolutely no idea' what the brand's core principles were based on the immersive content; this type of sentiment did not emerge from the Shanghai-based respondents, suggesting that lower awareness of the brand made them more accepting of the more contemporary content.

Just four Scotland-based and six Shanghai-based respondents picked up on the sustainability messages embedded within the immersive content, suggesting these messages needed to be more explicit. A sub-theme of connection emerged, with one respondent commenting on the 'sense of unity between different nations and cultures', one noting how the content was 'connecting countries' cultures' and another describing the brand as a 'product of history' with 'protection of the country'. This finding aligns with Kapferer's (2012) Brand Identity Prism, suggesting the *Culture* element of this model can utilise authentic, collaborative and respectful cultural appreciation (Jones and Childs 2020) in introducing a brand to a new target audience, to build a *Relationship* between the brand and new customers, using *Reflection* of

TABLE 10.4

What Do You Feel Are the Core Principles of the Brand?

Comments on Immersive Content (Shanghai)

Comments on Immersive Content (Scotland)

the characteristics of the new customer and their culture within marketing and communication collateral and channels. In an era of Digital Visual Multiculture (Han 2019), the visual imagery of both traditional video and immersive content affords an opportunity to transmit understanding and drive engagement, and in respect of the immersive content produced for the AF project, it was pleasing to gain positive user feedback on the amalgamation of the Scottish and Chinese cultures, without any sense of negativity or appropriation present.

10.3.3 Comparing Key Word Associations

After viewing the traditional HTH video content, respondents were presented with a set of nineteen words derived from the literature reviewed (Kapferer 2012; Keller 2019; Kim and Oh's (2020) sustainable fashion key word associations). As can be seen in Table 10.5, the most popular key words associated with the brand by Shanghai-based respondents were 'handmade' (90%), 'quality' (81%) and 'traditional' (65%). Similarly, the Scotland-based respondents selected 'traditional' (84%), 'quality' (79%) and both 'handmade' and 'heritage' (76% each), evidencing strong similarity in the associations made by both sets of respondents. However, fewer (48%) Shanghai-based respondents selected 'heritage', suggesting that the understanding of heritage is dependent on cultural 'cohort values' (Flynn and Foster 2009, 107).

After viewing the immersive content, changes in the respondents' brand associations were evident (see Table 10.6). The most popular key words now associated with the brand by Shanghai-based respondents were 'fashionable' (58%), 'futuristic' (52%)

TABLE 10.5
Words Associated with the Brand After Watching the Traditional Video Content

	Traditional	Quality	High-Quality	Heritage	Handmade	Natural	Sustainable
Scotland	32 (84%)	30 (79%)	28 (74%)	29 (76%)	29 (76%)	24 (63%)	20 (53%)
Shanghai	34 (65%)	42 (81%)	37 (71%)	25 (48%)	47 (90%)	31 (60%)	28 (54%)
Total	66 (73%)	72 (80%)	65 (72%)	54 (60%)	76 (84%)	55 (61%)	48 (53%)

TABLE 10.6
Words Associated with the Brand After Watching the Immersive Content

	Futuristic	Modern	High-Quality	Trendy	Heritage	Traditional	Fashionable
Scotland	26 (74%)	15 (43%)	15 (43%)	15 (43%)	14 (40%)	14 (40%)	13 (37%)
Shanghai	27 (52%)	19 (37%)	16 (31%)	10 (19%)	0	19 (37%)	30 (58%)
Total	53 (59%)	34 (38%)	31 (34%)	25 (28%)	14 (17%)	33 (37%)	43 (48%)

and both 'modern' and 'traditional' (37% each). The Scotland-based respondents selected 'futuristic' (74%) as the key association, followed by 'modern', trendy' and 'high-quality' (43% each). Given that neither the traditional video or the immersive content contained fashion-based garments or styles and showed only brief glimpses of the Harris Tweed® cloth, these results suggest the respondents' perceptions of the brand changed due to the immersive experience. This may, however, be due to a common, 'culturally-constituted' (Jung et al. 2021, 518) understanding of the role of virtual reality, depicted within socio-cultural discourse as a futuristic form of communication, rather than the specific content. Nevertheless, this is still an important finding as heritage brands like Harris Tweed® can find attracting new customers from younger demographics challenging (Keller 1999); immersive technologies may be a new way for these traditional brands to resonate (Keller 2019) with contemporary consumers.

Having watched the traditional video, 63% of Scotland-based respondents and 60% of Shanghai-based respondents associated the Harris Tweed® brand with the key word 'natural'. Similarly, 53% of the Scotland-based and 54% of the Shanghai-based respondents selected 'sustainable'. Disappointingly, after viewing the immersive content, the Scotland-based respondents showed reduced brand association with the words 'natural' (23%) and 'sustainable' (31%). Similar results were seen within the Shanghai-based respondents; natural (23%) and 'sustainable' (38%). It may be that the brightly coloured and synthetic, highly stylized aesthetic of the immersive content (see Figure 10.1) overwhelmed the incorporated sustainability messaging. Given the importance of sustainability in the fashion and textile sector (Environmental Audit Committee 2019), these findings suggest brand messaging around sustainability should be more explicit in both traditional and contemporary messaging media, aligning with the UN's (2023) Fashion Communications Playbook for consumer-facing communications. The findings also highlight the importance of congruence in multi-media content, where colour, text, font style and imagery combine to deliver successful messaging.

10.3.4 Comparing Comprehension of the Target Customer

To address the *Reflection* element of Kapferer's (2012) Brand Identity Prism (i.e. how a brand portrays its target audience), respondents were asked to describe the type of person the Harris Tweed Hebrides brand was targeting. After viewing the traditional video, the Shanghai-based respondents described a target customer that was over thirty, 'mature', 'middle-class', 'wealthy' and focused on 'quality'. This aligns with the views of the Scotland-based respondents, who also identified the target customer as 'older' or 'middle-aged', 'wealthy' and focused on 'quality' (see Table 10.7).

After viewing the immersive content, both sets of respondents perceived the target customer as 'younger' and 'fashionable', but still focused on 'quality' (Table 10.8). Although the immersive content elicited fewer themes from the respondents, the level of certainty that the brand was now deemed relevant to a younger customer would seem like a desirable outcome for the brand, with the immersive *Imagery* (level 2 on Keller's (2019) CBBE pyramid) providing context that these contemporary fashion consumers found relevant.

TABLE 10.7

What Type of Person Do You Think This Brand Is Trying to Target with This Content?

Comments on Traditional Content (Shanghai)

Comments on Traditional Content (Scotland)

TABLE 10.8

What Type of Person Do You Think This Brand Is Trying to Target with This Content?

Comments on Immersive Content (Shanghai)

Comments on Immersive Content (Scotland)

One UK respondent identified the target customer as 'gamers', with several other respondents describing the target customer as 'techy' or 'tech-savvy'. Similarly, a Chinese respondent described the target customer as 'people who are interested in AR and like to explore new things'. This does suggest that the use of VR and AR is still niche, not yet part of the wider fashion consumer experience. Interestingly, one

Scotland-based respondent speculated that 'possibly the Chinese consumer is more used to immersive content', evidencing some cultural stereotyping.

10.4 CONCLUSION

Overall, the findings show considerable alignment from the sampled respondents despite the difference in location and cohort (historical) values. This suggests that being part of the same generational cohort (i.e. Generation Z) is the overarching influence on attitudes among these contemporary fashion consumers. The findings are positive for traditional or heritage fashion and textile brands who may be considering adoption of immersive technologies to deliver their brand and marketing content, as they indicate this would be a way to engage with contemporary global fashion consumers who may not otherwise see these traditional brands as being of relevance. When using immersive technologies as an entry point into new markets, a collaborative ethos of cultural appreciation may be beneficial to brands as they develop visual content in an era of digital visual multiculture. Given brands such as Harris Tweed® have authentic sustainability in their supply chain and production processes, raising awareness of this type of brand as an alternative to unsustainable fast fashion is of huge importance; however, sustainability messages need to be explicit, with care taken that the immersive experience does not occlude important content and context. Finally, as many respondents were new to VR, lack of experience with this medium may have affected the opinions given. The still-niche use of VR headsets suggests focusing on more accessible immersive platforms such as mobile-enabled AR would be a safer entry point for fashion businesses at this time. In terms of the research limitations, the resource required to provide access to VR headsets and facilitate their use limited the sample size, meaning the results gathered are not generalisable. The survey focused on consumer attitudes towards the content (traditional and immersive) but did not explore user experience in terms of using the VR headsets as a channel or tool; this aspect could be explored in future studies by allowing users to fully explore the virtual world (full immersion) instead of limiting them to a pre-recorded fly-through (semi-immersion).

REFERENCES

Barnard, M. 2014. *Fashion Theory: An Introduction*. Routledge.

Boyd, D. E. and Koles, B. 2019. "An Introduction to the Special Issue "Virtual Reality in Marketing": Definition, Theory and Practice." *Journal of Business Research* 100: 441–444.

Cross, K., Mesjar, L., Steed, J. and Jiang, Y. 2023. "Immersive Innovations for the Communication of Heritage, Handcraft and Sustainability." *International Journal of Fashion Design, Technology and Education*. https://doi.org/10.1080/17543266.2023.2277264

Cross, K., Steed, J. and Jiang, Y. 2021. "Harris Tweed: A Glocal Case Study." *Fashion, Style and Popular Culture* 8(4): 475–494. https://doi.org/10.1386/fspc_00102_1

Environmental Audit Committee. 2019 February 19. "Fixing Fashion: Clothing Consumption and Sustainability." *House of Commons*. https://publications.parliament.uk/pa/cm201719/cmselect/cmenvaud/1952/1952.pdf

Flynn, J. Z. and Foster, I. M. 2009. *Research Methods for the Fashion Industry*. Fairchild Books.

Francis, Y. and Hoefel, F. 2018. *'True Gen': Generation Z and Its Implications for Companies*. MacKinsey and Company. http://www.drthomaswu.com/uicmpaccsmac/Gen%20Z.pdf

Han, H. 2019. "Moving From Cultural Appropriation to Cultural Appreciation." *Art Education* 72: 8–13.

Hollebeek, L. D., Clark, M. K., Andreassen, T. W., Sigurdsson, V. and Smith, D. 2020. "Virtual Reality through the Customer Journey: Framework and Propositions." *Journal of Retailing and Consumer Services* 55: 102056.

Interaction Design Foundation – IxDF. 2023. "What is immersion in Extended Reality (XR)?" *Interaction Design Foundation – IxDF*. https://www.interaction-design.org/literature/topics/immersion

Jones, D. and Childs, M. 2020. "Blackface Sweaters and Pizza with Chopsticks: Problem Cases and Strategies for Converting Cultural Appropriation into Cultural Appreciation in the Fashion Industry." *Journal of Cultural Marketing Strategy* 5(2): 143–152.

Jung, J., Yu, J., Seo, Y. and Ko, E. 2021. "Consumer Experiences of Virtual Reality: Insight from VR Luxury Brand Fashion Shows." *Journal of Business Research* 130: 517–524.

Kapferer, J. N. ed. 2012. *The New Strategic Brand Management (Advanced Insights and Strategic Thinking)*, 5th ed. Kogan Page.

Keller, K. L. 1999. "Managing Brands for the Long Run: Brand Reinforcement and Revitalization Strategies." *California Management Review* 41(3): 102–124.

Keller, K. L. 2019. *Strategic Brand Management: Building, Measuring, and Managing Brand Equity*, 5th ed. Pearson Education.

Kim, G., Jin, B. and Shin, D. C. 2022. "Virtual Reality as a Promotional Tool for Small Independent Stores." *Journal of Retailing and Consumer Services* 64.

Kim, Y. and Oh, K. W. 2020. "Which Consumer Associations Can Build a Sustainable Fashion Brand Image? Evidence from Fast Fashion Brands." *Sustainability* 12(5): 1703. https://doi.org/10.3390/su12051703

Li, C. et al. 2024. "Impacts of Government Attention on Achieving Sustainable Development Goals: Evidence from China." *Geography and Sustainability*. https://doi.org/10.1016/j.geosus.2024.08.011

Marshall, M. N. 1996. "Sampling for Qualitative Research." *Family Practice* 13(6): 522–526. https://doi.org/10.1093/fampra/13.6.522

Martirosov, S., Bures, M. and Zitka, T. 2022. "Cyber Sickness in Low-Immersive, Semi-Immersive, and Fully Immersive Virtual Reality." *Virtual Reality* 26: 15–32.

Platman, L. 2011. *Harris Tweed from Land to Street*. Frances Lincoln.

Rogers, E. M. ed. 1983. *Diffusion of Innovations*, 3rd ed. The Free Press.

Serdari, T. 2018. "The Carloway Mill Harris Tweed: Tradition-Based Innovation for a Sustainable Future." In *Sustainable Luxury, Entrepreneurship, and Innovation*, edited by M. Gardetti and S. Muthu. Springer.

Sherman, W. R. and Craig, A. B. 2018. *Understanding Virtual Reality*, 2nd ed. Morgan Kaufmann.

Slater, M. and Wilbur, S. 1997. "A Framework for Immersive Virtual Environments (FIVE): Speculations on the Role of Presence in Virtual Environments." *Presence, Teleoperators and Virtual Environments* 6: 603–616.

Steed, J., Cross, K. and Wilson, B. 2023. "Collaborating with a Scottish Heritage Brand Towards Enhancing and Preserving Sustainable Artisan Hand-Weaving Practices through a Knowledge Transfer Partnership." *Journal of Textile Design Research and Practice* 11(1–2): 127–147.

United Nations. 2015. "Sustainable Development Goals." *United Nations*. https://www.un.org/sustainabledevelopment/sustainable-development-goals/

United Nations Environment Programme. 2023. "The Sustainable Fashion Communication Playbook." *United Nations*. https://wedocs.unep.org/bitstream/handle/20.500.11822/42819/sustainable_fashion_communication_playbook.pdf?sequence=3

Wilson, R. H. B., Macleod, M. A. and Hogarth, M. 2021. "Sustainability and Heritage in Practice; The Case of Harris Tweed Hebrides." In *Firms in the Fashion Industry*, edited by L. Rienda, L. Ruiz-Fernandez and L. D. Carey. Palgrave Macmillan.

11 Understanding the Anatomy of Experience for Critical Analysis of Fashion Interactions

Erika Aline Rox

11.1 INTRODUCTION: ORIGINS OF EXPERIENCE

Experience is a multifaceted concept encompassing various dimensions and levels of interaction. Its etymology traces back to Latin, connoting 'testing possibilities', 'participation in events', or the acquisition of 'a skill gained by practice' (Merriam-Webster, n.d.a). Compared to those definitions, the German language offers a clearer distinction between the different aspects of experience. According to the Cambridge Dictionary (2014), '*das Erlebnis*' refers to a singular event that directly impacts an individual, emphasizing its emotional or sensory aspects. In contrast, '*die Erfahrung*' encompasses knowledge, wisdom, or skill gained over time through practice, indicating deeper learning and knowledge assessment.

Figure 11.1 illustrates how both concepts of experience align with the cognitive learning curve (Ebbinghaus 1913; Bills 1934), often also referenced as Wright's experience curve (Wright 1936), and the innovation adoption curve (Rogers 1983). This relationship highlights that the time humans require to evaluate new changes can either result in the formation of solid experiences or a drop of interest. In fashion, for instance, product and innovation adoption depends on first experiences for sustaining user interest and avoiding discontinuation along the adoption curve. When a single *Erlebniss* affects a given customer, positively or negatively, it inherently leads to a transformation in the way that individual perceives the world. Then, when fully evaluated, this perceptual shift can lead to a learned experience, or *Erfahrung* (Rox 2023). According to Malcolm Gladwell's *The Tipping Point* (2002), this process is normally driven by key individuals who play a big role in spreading trends and influencing others. These early adopters, by sharing their experiences, can help push a fashion innovation from the early stages of adoption to a tipping point, where it gains popularity and moves to a majority of consumers along the adoption curve.

Disregarding the factors influencing a consumer, both the first experience and the ability to process change are significant in determining the acceptance and retention of fashion products, trends, or brands. While a successful first contact is key for

DOI: 10.1201/9781003594482-14

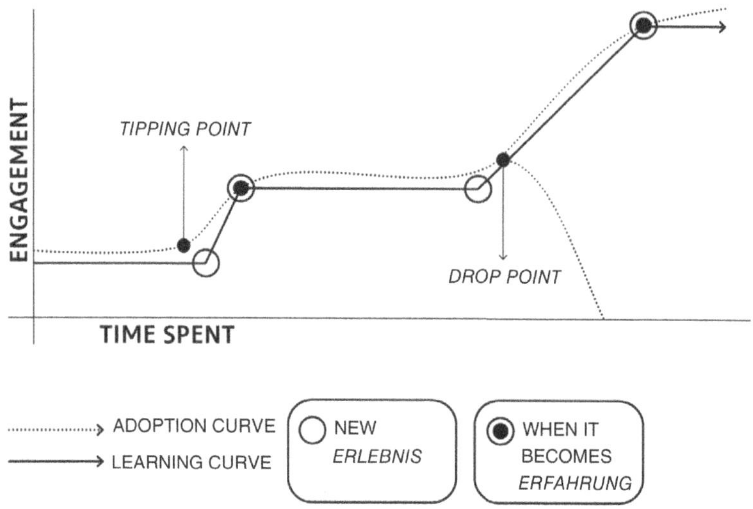

FIGURE 11.1 Comparing experience concepts through learning and adoption curves.

Source: adapted from Rox 2023; based on Ebbinghaus 1913, Rogers 1983 and Gladwell 2002)

embracing innovative concepts, a poor experience can hinder adoption, regardless of whether it occurs during a single interaction or through repeated engagement. Although fashion thrives on novelty, a product that fails to deliver a positive experience is unlikely to gain or sustain audience acceptance, potentially falling off the adoption curve at any stage.

Even though fashion, as a phenomenon, commonly serves as a social device for novelty (Lipovetsky 2009), there is a notable absence of innovative initiatives within the fashion system. Despite a conceptual alignment of designers and researchers with Industry 4.0, manufacturing processes and value chains related to clothing and accessories – the fashion milestones – remain tied to the Second Industrial Revolution paradigm (Bertola & Teunissen 2018). However, the emergence of Society 5.0 – a concept aiming to integrate physical and virtual realms to address societal and environmental challenges – reshapes the importance of Industry 4.0 and 5.0 technologies, now prioritized for citizen-centric benefits, representing a focus shift from manufacturing systems to individual perspectives (European Commission 2021). This brings up new opportunities for fashion brands and designers to focus on customized interactive experiences rather than unsustainable low-quality trends.

Drawing from existing literature, discussions, and personal insights, this paper presents the final synthesis of multiple combinations of conceptual frameworks aimed at systemizing the dynamics of subjective experiences. By capturing the essence of an agent's engagement with interactive systems and their emotional responses, the Anatomy of Experience framework seeks to provide a comprehensive understanding for analyzing fashion as experiences.

11.2 METHODOLOGY

The article begins with a literature review of some of the concepts that permeate the realm of experiences and their relevance to design and fashion systems. This process utilized academic search engines like Science Direct, Research Gate, and Google Scholar, as well as industry publications such as Vogue Business, Business of Fashion, Nielsen Norman Group, and Interaction Design Foundation, employing keywords such as 'experience', 'human-centered design', 'user experience', 'interaction design', 'digital design', and 'fashion'. Selection criteria included language, accessibility, and disciplinary relevance, with additional literature identified through the screening of previously selected research.

From the gathered findings, specific concepts and frameworks from the fields of Human Centered Design (HCD), User Experience (UX), User Interface (UI), and Interaction Design (IxD) were identified for further analysis and integration into the Anatomy of Experience framework. Distinguished by its graphical representation and the portrayal of dynamic interrelations among components of human experiences, this framework dissects each component to identify opportunities for design enhancement, primarily revolving around fashion perspectives rather than merely focusing on their original domains.

11.3 FASHION AS EXPERIENCE NARRATIVES

There is a deeply human aspect in the connection of garments as the closest interface between the body and the exterior world, especially mediating sensory experiences like touch, warmth, and even smell (Seymour 2008; Reilly 2014). Fashion, however, goes beyond the pure construction of garment pieces. Lipovetsky (2009) argues that fashion, without a fixed content or object, acts as a temporal social construct that blends together different aspects of collective life, such as art, architecture, culture, and ideals. Even within clothing systems, fashion experiences encompass multiple facets, from the act of wearing a garment or accessory to the allure for iconic pieces seen in fashion shows or communication campaigns. The seamless shopping experience for the perfect item, for instance, also reflects a time-engaging multifaceted nature of the fashion phenomena. Furthermore, in today's digitally infused world, technology integration enhances the propagation and accessibility of fashion as interactive spaces (Rox 2023). The convergence of *fashion and tech* to design smart products and services, bridging physical and virtual realms, introduces a new ecosystem connecting companies, users, and environments. Harnessing Industry 4.0 and 5.0 technologies, such as IT, Internet of Things, robots, artificial intelligence (AI), and virtual/augmented reality (VR/AR), offers opportunities for the fashion system to align product development with market demand, lifecycle transparency, and user needs within blended forms of self-expression (Bertola & Teunissen 2018; European Commission 2021).

As stated by Benyon (2014), interaction design 'is about designing interactive systems, products, spaces and services for people'. Given the prevailing physical nature of those people, there is a growing emphasis on enhancing the embodiment performance and sustainability of their clothing. Simultaneously, the ephemeral essence of fashion can still flow within extended layers of the same system – the extended realities (XR). As the initial buzz around speculative digital trends fades in traditional fashion circles for the past years, the presence of big fashion players in the gaming industry has become increasingly established. In an article for Vogue Business, Kotsoni (2024) gathers insights from digital fashion leaders about the future of the field, in which a good takeaway is that many consumers either struggle to understand digital fashion or feel overwhelmed by so many digital touchpoints, favoring tangible and offline assets.

This back-and-forth evolution presents a duality of challenges and opportunities for designers to navigate evolving interfaces and scenarios in the digital era. For this reason, understanding the dynamics of user experiences becomes crucial to ensure relevance and compliance with future scenarios.

11.3.1 The Experience System

According to Hassenzahl (2011), an experience is a narrative shaped by an individual's interactions with their surrounding world. It involves perception, motivation, and cognition, forming a complex connection of feelings, thoughts, and (inter) actions. For Benyon (2014), experiences are the result of the PACT system, i.e. *People* interacting with *Activities*, *Contexts*, and *Technologies*. In one interactive environment, numerous PACTs unfold simultaneously, prompting individuals to transition between those systems based on the inspiration of engagement, pleasure, and significance that derives from the experience at physical, behavioral, and social levels. Moreover, Gentile et al. (2007) offer a distinct perspective by outlining a perfect balance between the utilitarian and hedonic values of an experience. They identified functionality and usability as utilitarian values, while sensorial, emotional, cognitive, pragmatic, lifestyle, and relational aspects as pleasing components of user experience.

Overall, the experience system is holistic in its subjective meaning, and, as argued by McCarthy & Wright (cited in Benyon 2014), it must be conceived as a whole. However, in order to understand how it operates, the next chapter presents the Anatomy of Experience framework and analyzes each of its components.

11.4 THE ANATOMY OF EXPERIENCE

For constructing a comprehensible framework, components from the literature were mapped and structured for a coherent interpretation. Figure 11.2 shows some of the first design drafts of the tool.

The final version of the proposed framework is presented in Figure 11.3 and consists of the following key components: one agent, the interactive system, engagement, and emotional response. Each of these elements is further divided into smaller components and will be explored later in detail. Notably, the most designable aspect of the framework is the element of information, which permeates the entire system at varying levels of complexity. Enhancing the quality of information within the Anatomy of Experience is crucial for designers seeking to improve the overall user experience.

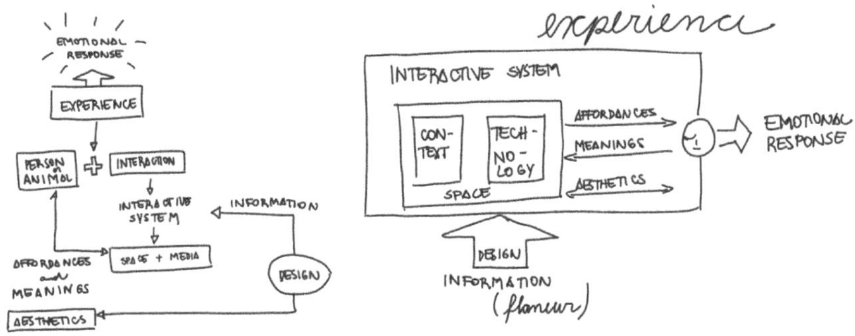

FIGURE 11.2 Sketching the Anatomy of Experience framework: first drafts.

Source: Rox 2023

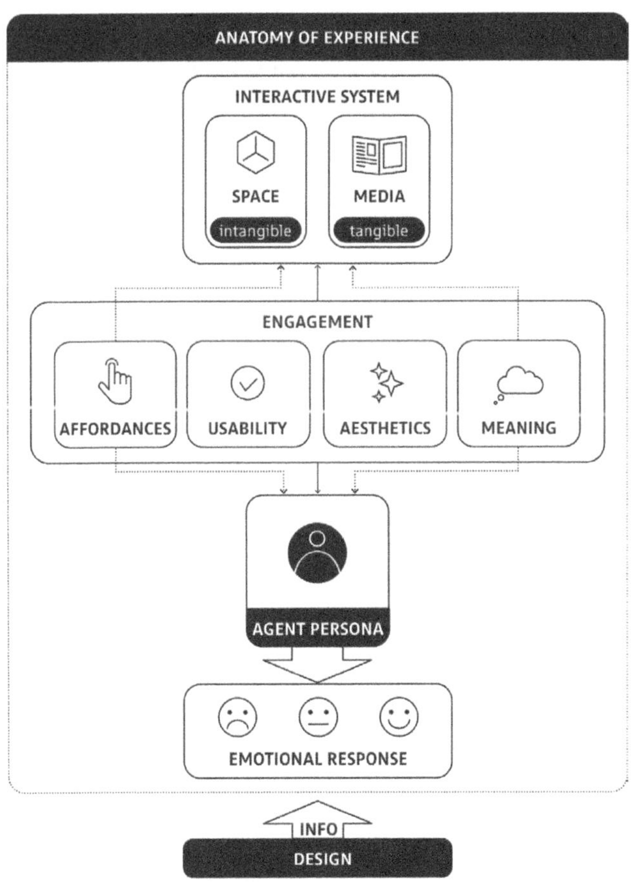

FIGURE 11.3 The Anatomy of Experience.

Source: Rox 2023

The following sections will analyze the anatomic components and explore how designers can effectively work with them.

11.4.1 Agent Persona

First of all, the experience invariably centers around an agent. Fashion specifically, as a social phenomena, is shaped and driven by people. Therefore, it is important to identify the appropriate target persona and gather authentic data and insights into the audience's behavior and mindset. This allows the proposal of valuable experiences being tailored to meet the agent's needs. For instance, drawing from Maslow's pyramid theory of human needs (as seen in McLeod 2025), a single product can address both the physiological need for protection and a social sense of belonging. However, envisioning how to effectively accomplish this is only possible through research. Donald Norman (2013), regarded many times as the father of UX, defends all forms of human research, ranging from quantitative surveys to qualitative interviews and observations. According to him, designers need both 'deep insights on real needs from a tiny set of people, versus broad, reliable purchasing data from a wide range and large number of people'. Moreover, as highlighted by fashion tech pioneer Sabine Seymour (2018), achieving this can and should involve the responsible utilization of big data while remaining mindful of the humans behind the data.

11.4.2 Interactive System

Secondly, any system designed for people is interactive. Interaction design, consequently, focuses on creating the media through which individuals engage with digital content – be it by producing, perceiving, playing with, consuming, manipulating, or communicating it (Benyon 2014). Interactive systems can exist without direct interaction, but its value significantly increases when it inspires some kind of engagement. While fashion systems may not always offer the most inclusive experiences, they are no exception to this principle (Rox 2023). To design for better experiences, it is important to consider the interactive system as a close relationship between two operable elements: space/s and media.

11.4.2.1 Space

Space plays a crucial role in shaping human experiences, serving as the backdrop for interactions and providing the environment within which experience unfolds. While intangible in nature, spaces are filled with tangible elements that can be intentionally designed to shape perception. Dourish and Bell (2007), for example, view space as 'an infrastructure through which we experience the world', which can be organized for practical actions and cultural meaning. This experiential interpretation of space acknowledges the interrelation of infrastructures, everyday life, and the technological, social, and cultural aspects of experience.

Professor David Benyon's work in his book *Spaces of Interaction, Places for Experience* (2014) is fundamental in this area. Benyon explores various types of spaces, including physical, digital/cyber, and blended spaces, as well as information, conceptual, navigation, and social spaces. The professor also identifies four crucial aspects to consider when conceptualizing a space: Ontology, Topology, Volatility, and Agency. Ontology refers to the distribution and function of elements in a space, while Topology considers how elements relate to each other within the space. Volatility relates to the dynamics of moving and changing elements over time, and Agency concerns the agent personas – the first component discussed in this study – and artificial agents of action.

Similarly, Hansen (2006) examines physical, virtual, and mixed realities as frameworks for embodying human experiences. The author argues that 'all reality is mixed reality', emphasizing the ontological function of embodiment and the role of self-movement as the tactile and sensory realities of perception. Hansen also introduces the concept of 'wearable space', suggesting that a space becomes wearable when embodied affectivity drives spatial operations. At one end, there is a pragmatic, technical outlook, envisioning a fluidity between the human body and a computer – an extension of the body leveraging data and internal processes; at the other end lies an abstract projection of a mutual fluidity between body and space.

In recent years, fashion, technology, and creative sectors have centered on these three fundamental spaces: physical (referred to as In Real Life or IRL), digital (Uniform Resource Locator or URL), and the emerging concept of 'phygital' space. Despite the growing influence of technology leading us towards the exploration of phygital spaces, it is noteworthy that the significance of physical space persists as long as human experiences are tied to living, embodied existences.

11.4.2.2 Media

Media, the plural form of a medium, is intricately linked to the forming components of Benyon's (2014) PACT framework, context and technology. The first refers to the circumstances that prompt a message to be conveyed, while technology encompasses the tools and methods employed for communication. According to Shaleph O'Neil (2008), a medium serves as a means or vehicle of self-expression and communication with others, and 'has to be some kind of physical entity'. This highlights the fundamental role of the body in every act of mediation, as it interacts with the material world around it. Essentially, the body and how it is dressed is indispensable for mediating agency.

Media technologies carry ideas, providing a platform for expression and offering content for contemplation, reflection, and utilization. Examples range from traditional materials such as paper, wood, and textiles to modern electronic devices produced on a mass scale. O'Neil portrays the emergence of new media and technologies as big transformers of interaction dynamics, allowing individuals to enhance their capabilities through the creation and consumption of innovative content.

Fashion media, including magazines, websites, and social media platforms, play a crucial role in disseminating trends, influencing consumer behavior, and shaping perceptions of beauty and style. In this sense, Lipovetsky (2009) argues that contemporary individualism and the rise of new media are interconnected. The abundance of content, he says, compels individuals to form opinions and compare themselves to others, leading to a diversification of values and a weakening of traditional forms of culture. O'Neil (2008) also states that fashion itself can be considered a form of media, conveying messages about identity, culture, and society. Through clothing and accessories, individuals communicate their personal style, values, and aspirations.

Overall, new media fosters the spread of knowledge, serving as a catalyst for individual and collective transformation. Just as traditional media channels evolve with advancements in technology, the fashion industry embraces new digital platforms and technologies to reach audiences and engage consumers in innovative ways.

11.4.3 Engagement

The advent of social media has transformed the way people interact with each other, enabling them to engage in digital spaces through intuitive information codes such as liking, sharing, and commenting. Benyon (2014) asserts that people, as social beings, naturally interact to share their thoughts and express their identity within a specific narrative. Due to social media, engagement is recognized as a value creator and, consequently, highly valued in the digital scope.

This component, in fact, encompasses various levels of involvement, with total immersion – known as the state of flow – being the deepest form (de Byl 2015). Flow occurs when an individual is completely absorbed in an activity, experiencing intense focus, involvement, and enjoyment (Csikszentmihalyi 1997), and is especially important for gamified experiences. Flow is also the state of total enchantment during a show, a film, or a book, or the complete concentration during a sport activity. Achieving it requires a meaningful narrative, clear goals, and incremental achievements. The gaming industry has mastered this concept, keeping players engaged as they strive to overcome challenges (Tan & Jansz 2008). From the deepest state of flow to more superficial levels of social media, interaction systems must strike a balance between four elements to enable a seamless engagement: affordance, meaning, aesthetics, and usability.

11.4.3.1 Affordance and Meaning

Affordance, a term introduced by psychologist J. J. Gibson, refers to the perceived actions enabled by an interactive system, reflecting an individual's perception of the world. Norman (2013) states that 'an affordance is a relationship between the properties of an object and the capabilities of the agent that determine just how the object could possibly be used.' Visual affordances, as noted by Norman (2007), play a crucial role in guiding instinctive behavior, especially in the context of new and technological artifacts. On a broader level, some affordances can be linked to Carl Jung's Archetypes, which are universal symbols embedded in the collective unconscious that share similar meanings across cultures and generations. Jung's (1936) theory suggests that the collective unconscious stores humanity's shared memories, primarily in the form of archetypes that manifest universally in the psyche. This collective understanding of archetypes may explain why individuals intuitively know how to navigate certain life situations even without personal experience.

In the realm of fashion, Norman's (2004) insights focus solely on the style, status, and meaning of clothing rather than its role for embodied activity. Spurrett and

Brancazio (2023), on the other hand, propose that clothing should be studied as an affordance-transforming technology as it modifies appearance and capabilities, shapes agency, and expands possibilities for action. The affordances provided by clothing are influenced by both the properties of the body and its environment; therefore, clothing can also enhance the functional performance of the body by providing, for example, thermal protection or functional features – like pockets. Moreover, as emphasized by Sabine Seymour (2008), garments serve as a direct interface between body and the external world, conveying emotions, experiences, and meanings. Consequently, garment manipulation often stems from the wearer's creativity, cultural influences, and desire for self-expression. Ultimately, Seymour depicts that the context determines the functional and expressive definition of fashionable wearables, underscoring the importance of aligning aesthetics with functionality and meaning.

11.4.3.2 Aesthetics and Usability

In the domain of fashion, aesthetics plays a central role as it is intricately connected to the creation of beauty. As defined by the Merriam-Webster dictionary (n.d.b), aesthetics is a branch of philosophy that explores the nature of beauty, art, and taste, as well as the creation and appreciation of beauty. The Greek word 'aisthesis', which relates to recognition through sensory perception, emphasizes that beauty encompasses sight, hearing, touch, smell, and taste, and is influenced by culture and cognition.

Designers with keen observation skills often have an intrinsic ability to uncover and shape new perspectives of aesthetic pleasure, frequently guided by established theoretical principles – such as the golden ratio, the rule of thirds, and Gestalt laws – as visual guidelines. However, as argued by Lim et al. (2007), interactive artifacts are significantly different from those typically addressed in traditional design disciplines. They require consideration of the unique dynamics, flexibility, and intelligence made possible by computing technologies. To address this, the researchers developed a tool called 'the interaction gestalt', which must be designed to evoke specific user experiences. In their words, 'The designer has to anticipate how a certain gestalt will be experienced by a user, and that anticipation has to be translated back into ideas on how the gestalt should be shaped.'

The aesthetics of interaction is a relatively recent concept; however, by abstracting aesthetic principles, designers can enhance the semantics, usability, and accessibility of any designed artifact (Rox 2023). In fact, the Aesthetic-Usability Effect, as mentioned by Kate Moran (2017), accentuates the importance of balancing design form and function. While visually pleasing designs may mask minor usability or accessibility issues, they may not effectively facilitate deeper interactions. Usability, on the other hand, evaluates the ease of interaction, and represents a fundamental but often undervalued aspect of design. Norman (2004) implies that fashion systems frequently apply usability principles in unconventional ways. For instance, some retail arrangements intentionally make it challenging for customers to purchase the most desired items while making impulse items easily accessible. The author defends that when stores implement correct strategies, the shopping experience can be emotionally positive for the shopper and financially profitable for the seller.

Hence, the phases of prototyping and testing are essential design processes that enable iterative improvement. In clothing design, prototyping is a phase for iteration not only to find the perfect fit or material but also to connect the product with users and test the wear experience (Rox 2023). As endorsed by Norman (2013), involving users in the testing process provides valuable insights into usability and aesthetic issues.

11.4.4 Emotional Response

The last component of the Anatomy of Experience is a direct consequence of the overall interaction – the emotional response. Norman's work delves into the intricacies of human cognition, exploring how emotions, moods, traits, and personality shape an individual's interactions with the world. In his book *Emotional Design* (2004), the author identifies three levels of emotional responses, as illustrated in Figure 11.4: visceral (first impression), behavioral (during use), and reflective (memory). While the visceral and

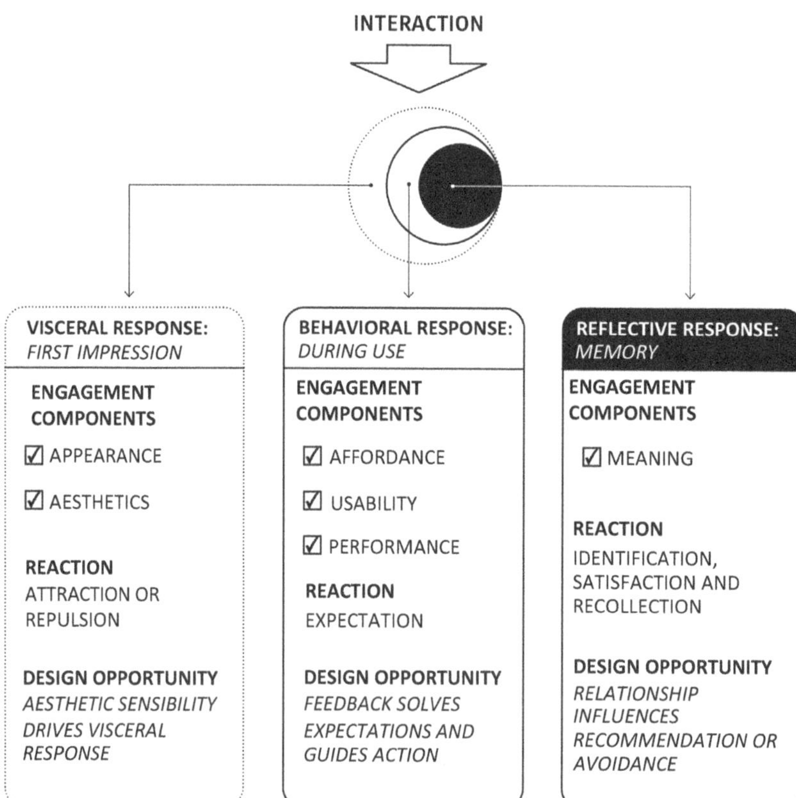

FIGURE 11.4 The three levels of emotional response according to Don Norman.

Source: adapted from Rox (2023)

behavioral levels involve immediate affective and judgmental responses, the reflective level allows for interpretation and understanding influenced by cultural, experiential, educational, and individual differences. In other words, while emotions shape short-term behavior, moods have a longer-lasting impact. On an even longer term, traits and personality, though enduring, remain adaptable over time. In *The Design of Everyday Things* (2013), he further explores the inseparable connection between cognition and emotion. In his words, 'affect and cognition are information-processing systems', concluding that cognitive thoughts lead to emotions, and vice-versa, as the brain interacts with the world.

Paradoxically, Daniel Goleman (2011) states that despite a sense of compulsive bliss, the flow state of total immersion remains free from emotional interference. While challenging activities typically demand increased cortical stimulation, flow occurs at the peak of mental capacity with minimal effort, making high performance feel effortless. It does not always engage all the senses but instead relies on imagination and the pleasure of it – often an even more important factor for creating seamless experiences. Norman always highlights the importance of designing for different emotional levels and qualities, acknowledging the complexity of evoking both positive and negative emotions simultaneously. In this matter, Desmet and Fokkinga (2013) tackle the challenge of evoking contrasting emotions by proposing a framework of 10 emotional qualities. This framework aims to foster rich emotional experiences by shifting perceptions and attitudes across different emotional levels, ultimately making experiences more memorable.

Similarly, the Interaction Design Foundation (2024) states that emotion-driven designs go beyond meeting user needs, also enriching the overall experience. However, as de Byl (2015) highlights from the context of integrating emotions into game interactivity, understanding user emotions involves more than measuring bodily response or creating non-players narratives. It requires embedding emotion recognition and generation into design, applying insights from psychology and neuroscience, utilizing emotion-sensing technology, and mastering affect expression in characters and cognitive models of emotion. Both de Byl and the foundation support their claims by referencing psychologist Robert Plutchik, who developed the Wheel of Emotions – a color-based visual tool – that categorizes opposing emotions and their varying intensities into primary, secondary, and combined levels. This tool enables designers to visualize contrasting emotions and make informed decisions to shape rich and immersive experiences.

Thus, understanding and leveraging emotions is fundamental for all areas of design. Historically, fashion has effectively harnessed emotions to create desirability – though not always ethically. As Norman (2004) suggests, despite constant perceptions of manipulation, consumers willingly engage in the dynamic of fashion, driven by strong emotional connections and the allure of self-fulfilling promises.

11.4.5 The Designer as 'the Information Flaneur'

As previously discussed, information is a crucial tool for the human experience. The concept of 'the information flaneur', as outlined by Dörk et al. (2011), offers a fresh perspective on information seeking, drawing inspiration from the urban flaneur found in literature. Just as the character leisurely interprets and reimagines the cityscape, designers navigate information spaces to uncover beauty and significance. By adopting this curious, creative, and critical approach, designers can broaden their focus from mere problem-solving to the creation of rich experiences. Like the information flaneur, designers rely on human-centered perspectives to comprehend and address the cognitive, perceptual, and affective dimensions of design. In summary, designers can shape interactions and enhance the overall user experience through intuitive, meaningful, and aesthetically pleasing design choices based on observational insights.

11.5 CONCLUSION

In recent years, digital technologies have significantly transformed all industries, presenting both challenges and opportunities for creating immersive and personalized experiences. The adoption of new digital platforms has revolutionized how fashion is consumed and engaged with. However, amidst this digital evolution, the importance of physical spaces and embodiment remains significant. While technological advancements aim to enhance consumption and well-being, issues such as sustainability and accessibility require careful consideration.

As noted by Bill Buxton (2007), 'design is about the future.' Looking ahead, interaction design plays a crucial role in shaping fashion experiences by blending digital and physical interactions. Emotion-driven design and engagement strategies offer valuable insights into proposing immersive and meaningful experiences. By applying the concepts discussed with the Anatomy of Experience to prioritize user experiences and incorporate human-centered perspectives into their projects, designers can continue to enrich lives, foster emotional connections, and elevate the quality of design solutions for generations to come.

11.6 ACKNOWLEDGMENTS

I would like to thank Professor Paola Bertola and Angélica Vandi for their guidance during my master's thesis at Politecnico di Milano, where *The Anatomy of Experience* was first developed, and for the Fellowship opportunity at the same institution that led to this article. I also acknowledge ChatGPT for the help at refining specific wording and translations, as I believe that new technologies can be thoughtfully used to support our development process.

REFERENCES

Benyon, David. 2014. "Spaces of Interaction, Places for Experience." In *Synthesis Lectures on Human-Centered Informatics*. Morgan & Claypool Publishers. https://doi.org/10.1007/978-3-031-02206-7.

Bertola, Paola, and José Teunissen. 2018. "Fashion 4.0: Innovating Fashion Industry through Digital Transformation." *Research Journal of Textile and Apparel* 22, no. 4: 352–369. https://doi.org/10.1108/RJTA-03-2018-0023.

Bills, Arthur G. 1934. *General Experimental Psychology*. New York: Longmans, Green and Co. https://doi.org/10.1037/11314-000.

Buxton, Bill. 2007. *Sketching User Experience: Getting the Design Right and the Right Design.* San Francisco: Morgan Kaufmann Publishers.

Cambridge Dictionary. 2014. "Translation of Experience." *English–German Dictionary.* https://dictionary.cambridge.org/dictionary/english-german/experience [Accessed April 12, 2024].

Csikszentmihalyi, Mihaly. 1997. *Flow and the Psychology of Discovery and Invention.* New York: Harper Perennial.

de Byl, Penny. 2015. "A Conceptual Affective Design Framework for the Use of Emotions in Computer Game Design." *Cyberpsychology: Journal of Psychosocial Research on Cyberspace* 9, no. 3, Article 4. https://doi.org/10.5817/CP2015-3-4.

Desmet, Pieter and Fokkinga, Steven. 2013. *Ten Ways to Design for Disgust, Sadness, and Other Enjoyments: A Design Approach to Enrich Product Experiences with Negative Emotions.* International Journal of Design. 7. 19–36.

Dörk, Marian and Carpendale, Sheelagh and Williamson, Carey. 2011. "The Information Flaneur: A Fresh Look at Information Seeking." In *Proceedings of the SIGCHI Conference on Human Factors in Computing Systems*, 1215–1224. New York: Association for Computing Machinery. https://doi.org/10.1145/1978942.1979124.

Dourish, Paul, and Genevieve Bell. 2007. "The Infrastructure of Experience and the Experience of Infrastructure: Meaning and Structure in Everyday Encounters with Space." *Environment and Planning B: Planning and Design* 34, no. 3: 414–430. https://doi.org/10.1068/b32035t.

Ebbinghaus, Hermann. 1913. *Memory: A Contribution to Experimental Psychology.* New York: Teacher's College, Columbia University.

European Commission, Directorate-General for Research and Innovation, Breque, Manuel, Laurent De Nul, and Alexandros Petridis. 2021. *Industry 5.0: Towards a Sustainable, Human-Centric and Resilient European Industry.* Publications Office of the European Union. https://data.europa.eu/doi/10.2777/308407.

Gentile, Chiara and Spiller, Nicola and Noci, Giuliano. 2007. "How to Sustain the Customer Experience: An Overview of Experience Components that Co-create Value With the Customer." *European Management Journal* 25, no. 5: 395–410. Elsevier Ltd https://doi.org/10.1016/j.emj.2007.08.005

Gladwell, Malcolm. 2002. *The Tipping Point: How Little Things Can Make a Big Difference.* Boston: Back Bay Books.

Goleman, Daniel. 2011. *Inteligência Emocional* [Emotional Intelligence]. Translated by Marcos Santarrita. Rio de Janeiro: Objetiva

Hansen, Mark B. N. 2006. *Bodies in Code: Interfaces with Digital Media.* New York: Routledge. https://doi.org/10.4324/9780203942390.

Hassenzahl, Marc. 2011. *User Experience and Experience Design.* https://www.researchgate.net/publication/259823352_User_Experience_and_Experience_Design

Interaction Design Foundation. 2024. "Putting Some Emotion into Your Design – Plutchik's Wheel of Emotions." *Interaction Design Foundation – IxDF.* March 2. https://www.interaction-design.org/literature/article/putting-some-emotion-into-your-design-plutchiks-wheel-of-emotions.

Jung, Carl G. 1936. "The Concept of the Collective Unconscious." *In Collected Works* 9, no. 1: 42.

Kotsoni, Elektra. 2024. "Is Digital Fashion Still in Fashion?" *Vogue Business.* https://www.voguebusiness.com/story/fashion/is-digital-fashion-still-in-fashion.

Lim, Youn-kyung and Stolterman, Erik and Jung, Heekyoung and Donaldson, Justin. 2007. "Interaction Gestalt and the Design of Aesthetic Interactions." In *Proceedings of the 2007 Conference on Designing Pleasurable Products and Interfaces (DPPI '07)*, 239–254. New York: Association for Computing Machinery. https://doi.org/10.1145/1314161.1314183.

Lipovetsky, Gilles. 2009. *O império do efêmero: A moda e seu destino nas sociedades modernas* [The Empire of Fashion: Dressing Modern Democracy]. Translated by Maria Lucia Machado. São Paulo: Companhia das Letras.

McLeod, Saul. 2025. *Maslow's Hierarchy of Needs*. Simply Psychology. https://www.simplypsychology.org/maslow.html

Merriam-Webster. n.d.a "Aesthetic." *Merriam-Webster.com Dictionary*. https://www.merriam-webster.com/dictionary/aesthetic [Accessed April 26, 2024].

Merriam-Webster. n.d.b "Experience." *Merriam-Webster.com Dictionary*. https://www.merriam-webster.com/dictionary/experience [Accessed April 6, 2024].

Moran, Kate. 2017. "The Aesthetic-Usability Effect." *Nielsen Norman Group*. https://www.nngroup.com/articles/aesthetic-usability-effect/ [Accessed April 6, 2024].

Norman, Donald A. 2004. *Emotional Design*. New York: Basic Books.

Norman, Donald A. 2007. *The Design of Future Things*. New York: Basic Books.

Norman, Donald A. 2013. *The Design of Everyday Things*. New York: Basic Books.

O'Neil, Shaleph. 2008. *Interactive Media: The Semiotics of Embodied Interaction*. Springer. https://doi.org/10.1007/978-1-84800-036-0

Reilly, Lucy. 2014. "The Shift from 3D Body Scanned Data to the Physical World." Shapeshifting: A Conference on Transformative Paradigms of Fashion and Textile Design, 14–16 April 2014, Auckland, New Zealand. https://hdl.handle.net/10292/8564.

Rogers, Everett M. 1983. *Diffusion of Innovations*. 3rd ed. New York: The Free Press.

Rox, Erika A. 2023. "The X Files: Investigating Fashion Experiences." Master's thesis, Politecnico di Milano. https://doi.org/10.13140/RG.2.2.34633.35685.

Seymour, Sabine. 2008. *Fashionable Technology – The Intersection of Design, Fashion, Science, and Technology*. New York: Springer Wien.

Seymour, Sabine. 2018. "Our Second Skin – Meet Dr. Sabine Seymour, Founder of SUPA." Interview by Kavita Pradeep. HCVC on Medium. https://blog.hardwareclub.co/our-secondskin-meet-dr-sabine-seymour-founder-of-supa-45cbfdb3c0f [Accessed April 12, 2024].

Spurrett, David and Brancazio, Nicholas. 2023. "Fashioning Affordances: A Critical Approach to Clothing as an Affordance Transforming Technology." *Philosophical Psychology*. https://doi.org/10.1080/09515089.2023.2263478.

Tan, Ed S. and Jansz, Jeroen. 2008. *The game experience*. Product Experience. https://doi.org/10.1016/B978-008045089-6.50026-5

Wright, Theodore P. 1936. "Factors Affecting the Cost of Airplanes." *Journal of the Aeronautical Sciences* 3: 122–128. https://doi.org/10.2514/8.155.

12 Application of Extended Reality (XR) Technologies in the Fashion World

Md. Mahbubor Rahman, Jannatul Fardus, Naila Sharmin Kanta, Md. Mahmudul Hasan, Ghada Soliman, and Abu Sadat Muhammad Sayem

12.1 INTRODUCTION

The emergence of extended reality (XR) reflects the hyperconnected culture of today, which expresses a desire to immerse and broaden their senses in cutting-edge technological advancements (Silvestri, 2022). The advantages of extended reality are not being overlooked, and the fashion sector is gradually investing in technological innovations. According to analysts from McKinsey & Company and the Business of Fashion's research "The State of Fashion Technology," fashion brands are predicted to boost their technology investment by 3 to 3.5 percent by 2030 (Hillmann, 2021). Additionally, the worldwide XR market is expected to reach over 250 billion dollars by 2028, having reached 28 billion dollars in 2021, as reported by Statista (Amed et al., 2022). Extended reality opens multiple possibilities to boost brand value. This platform allows gathering detailed information on customers, such as their interactions, interests, and product usage, which can be utilised to improve marketing tactics and offer individualised experiences (Hackl & Wolfe, 2017). According to the standpoint of marketing, XR technology is being used in fashion shows, promotional activities, and commercial debuts. Multiple insightful applications improve consumer experience in both physical stores and online shopping, such as in-store navigation applications, magic mirrors, and virtual showrooms.

12.2 EXTENDED REALITY (XR)

All integrated real-and-virtual environments and human-machine interactions arising from computer technologies and wearables are designated to as XR, where the "X" represents a variable applicable to any existing or forthcoming spatial computing technologies. The term extended reality refers to a collection of immersive technologies (Figure 12.1), including virtual reality (VR), augmented reality (AR), and mixed reality (MR), which can create various real and virtual environments and engage users through auditory, visual and potentially tactile and olfactory stimuli (Li et al., 2022).

DOI: 10.1201/9781003594482-15

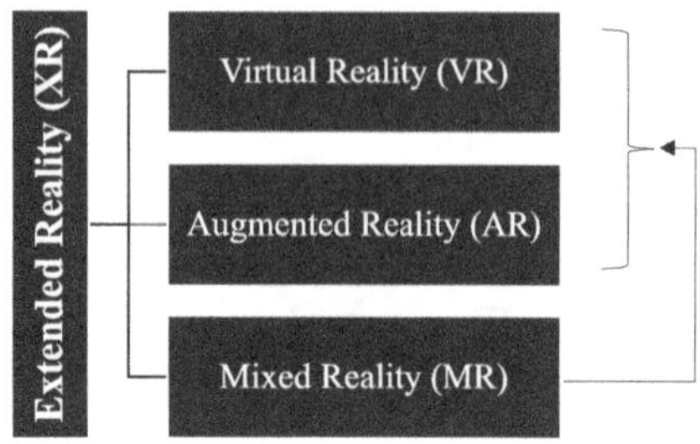

FIGURE 12.1 Technologies encompassed by extended reality.

Extended reality exemplifies how technological improvements across various domains drive innovation and provide new application possibilities, enabling numerous spectacular computer-generated experiences (Stanney et al., 2021). It includes all real-and-virtual integrated worlds and interactions between humans and machines developed through computer technology and wearable gadgets (Fast-Berglund et al., 2018). The predominant focus of XR projects has been on events such as marketing initiatives, presentations, and showcases. XR has been integrated into fashion shows, advertising, and product launches for marketing purposes, alongside numerous applications aimed at improving customer services, such as virtual outlets, smart mirrors, in-store location applications in both e-commerce and physical retail environments (Sabatini et al., 2023).

Today's XR technology can boost the customer's experience since recent advances in the technology are giving customers reliability (Jung et al., 2024). To support customers in making decisions during online purchasing, XR has been recommended to render particular consumers with realistic, immersive, 3D shopping experiences, including the ability to try on clothing and accessories as well as improving product or service details. By employing haptic devices, visual 3D images for product display, and audio-visual enhanced atmospheres, a realistic online store is emulated to give customers an active presence and enveloping scenario while precisely monitoring their virtual online shopping responses (Chen et al., 2024).

Extended reality (XR) technology could transform how advertising would be delivered and consumed. Despite some limitations in terms of cost, usability, and accessibility, XR advertising can boost consumer engagement, brand awareness, and purchase intent. Virtual product presentations, interactive experiences for brands, dynamic storytelling, and gamification are just a few ways that XR advertising can be implemented. Due to its distinctive attributes – such as the ability to deliver memorable, customised, and engaging experiences – XR holds significant potential as a resource for marketers aiming to attract customers and influence their behaviour.

Immersive technologies like XR have been demonstrated to be effective in increasing awareness and encouraging behavioural change concerning environmental and sustainability issues. Consumers can confront the real environmental effects of the fashion business through the visually enthralling depiction of some sustainability fashion challenges linked to production processes. By initiating an experience that encourages visual acknowledgement of the fashion industry's environmental impact and, consequently, a potential degree of consciousness of actions that could be implemented to mitigate the adverse impact, it can enhance users' comprehension of the sustainability exigency and encourage the demand for change (Spadoni et al., 2024). Online branding and e-commerce platform adoption are being promoted by the fashion industry's digital transformation process. The online fashion market is currently valued at 332 billion USD, 28 percent of all e-commerce businesses (Morotti et al., 2022). However, applying XR technologies in fashion settings is still restricted to exclusive and infrequent exhibitions to create a "wow-effect" that draws clients and promotes brand recognition. The cause of this is an accumulation of (a) XR technologies' immaturity, (b) consumers' poor acceptance, and (c) non-specialists' (consumers' or businesses') lack of comprehension of how to utilise them. Optimistically, global trends show that many consumers are becoming more receptive to fashion e-commerce despite early reluctance (Morotti et al., 2022).

12.3 EXTENDED REALITY (XR) TECHNOLOGIES IN FASHION

12.3.1 Virtual Reality (VR) in Fashion

Virtual reality is a computer-simulated, real-time visualisation that lets users interact with the world and become fully immersed in it. Since virtual reality generates a highly realistic computer-simulated world, it is one of the most recent technological developments with enormous potential for the fashion industry. Fashion retailers quickly experimented with virtual reality in their stores to attract customers. Fashion retailers like Tommy Hilfiger and Topshop install virtual reality runway shows in their retail shops to give customers a three-dimensional, front-row view of runway events. Additionally, J.C. Penney employed VR technology in their stores during the holiday season to captivate shoppers by providing distinctive experiences. These lifelike renderings of products and amenities will enhance customer emotion and store impression, ultimately resulting in increased sales (Park et al., 2018).

There are two categories of virtual reality systems: immersive and non-immersive. The primary classification criterion is the contrast between the output equipment and the effect the equipment produces. Non-immersive VR systems use infrared monitors, whereas head-mounted displays are used by immersive VR systems (Wu & Kim, 2022).

The provision of an immersive digital garment showcase within a captivating VR environment enables fashion developers and merchants to transcend the constraints of conventional garment creation and display. Designers can navigate the simulation of a virtual retail store, selecting from an array of avatar-garment combinations and examining apparel from multiple perspectives. CLO3D is a specialised software tool for fashion designers that facilitates the creation of 3D designs suited for actual production. Fashion designers can assess its benefits by examining the fitting of an avatar's clothes from many angles and movements, including intricate details such as drapes and wrinkles (Sarakatsanos et al., 2024).

Virtual reality impacts on fashion shows, platforms for showcasing clothing and connecting consumers with fashion firms. It overcomes zonal, cultural, and temporal boundaries and has distinct spatial qualities that cannot be felt in a physical space (Kim, 2023). VR fashion shows underscore audiences' satisfaction by enhancing their feelings of importance as they have the ability to remove problems associated with traditional fashion shows like buying tickets and managing a good seat. VR reflects positive feelings of ebullience, liberty, and freedom from social order and market institutions. However, the study also discovered that VR experiences can cause people to feel afraid, anxious, and alone, which makes them dislike VR technology. The mismatch between the realism of some elements of the experience, such as its vividness, and the absence of other elements, like the lack of social connections, cause anxiety in virtual reality (Jung et al., 2021). Virtual reality is also a helpful tool in garment design in a design classroom and a source of student inspiration. In addition to offering unthinkable inspiration sources that are not achievable in real life, virtual reality (VR) enabled the exploration of design ideas in a novel manner that sparked creativity (Starkey et al., 2021).

12.3.2 Augmented Reality (AR) in Fashion

Augmented reality can be explained as a direct or indirect scene of a physical, real-world circumstance that has been improved or supplemented by the inclusion of virtual, machine-generated data. AR integrates virtual and real-world objects and is interactive and 3D registered. AR technology aims to make a user's life easier by providing virtual information to both his immediate circumambient and any indirect scene of the real world, like a live video stream (Reality, 2011).

AR technology has been employed in various fashion-related applications, including Magic Mirrors in retail stores, 3D features in fashion magazines, augmented fashion shows using tagged models or animated holograms, and advertisements utilising slightly enlarged QR codes. The foundation of Magic Mirror technology is "flawless tracking," which enables the computer screen to operate as a mirror. Unlike conventional mirrors that merely reflect objects placed in front of them, and standard monitors, which display images that are not dependent on their surroundings, the display in this case functions as an interface that replicates what it captures through its digital camera and adds additional virtual contents to the image through algorithms that measure spatial coordinates and movements. For example, in a virtual fitting room, one can try on clothing by standing in front of a Magic Mirror and choosing from various models, colours, and sizes. The touchscreen interface will provide an image of you adorned in the specified garment, featuring the corresponding model, colour, and size (Logaldo, 2016).

The usage of AR in e-commerce raises the value of e-commerce in general, particularly when it comes to self-service and user customisation. Consumers can engage directly with the application to try them out and see how they appear when wearing those items. The most important factor for beauty products is how accurately the product's colour is displayed. Therefore, users rely on the augmented reality function to decide whether or not to purchase the product (Gabriel et al., 2023).

The enhanced visualisation through using AR technology not only boosts customer confidence but also makes returns less likely, making the shopping process easier and raising satisfaction levels all around. Additionally, AR's social sharing feature enables users to ask peers for advice, strengthening the influence of social networks on purchasing decisions. Using web-based platforms, AR technology might potentially work with product developers to review, illustrate, and personalise things through head-mounted displays. Product developers can make better decisions by employing AR since it allows them to foresee potential problems that might occur when using the physical output and come up with solutions for them (Reddy, 2024).

12.3.3 Mixed Reality (MR) in Fashion

Mixed reality involves immersing the user in an interactional environment that is either virtual with real-world augmentation (augmented virtuality) or real with virtual asset augmentation (augmented reality) (Hughes et al., 2005). It combines real-world components into a virtual environment or, alternately, virtual computer graphics items into a real 3D scene (Pan et al., 2006). Mixed reality application emphasise on virtual content without relating it to the physical world (Quint et al., 2015). Effective communication of prefabrication designs, site installation, and the planning and carrying out maintenance operations could all benefit from MR technologies (Dunston et al., 2017).

MR could now be implemented in the field of sales of relatively low-end, high-volume items, such as apparel, particularly sports equipment. The setup of an MR infrastructure within a store could potentially facilitate online visitors' access to the retail space. Mixed reality has the capability to create a synergy between retail channels that is comparable to the established advantages of multichannel marketing and advertising (Koontz & Gibson, 2002).

The economic ramifications of mixed reality in retail are significant, suggesting considerable market expansion and financial prospects for businesses. Nonetheless, implementation of MR poses hurdles, such as the necessity for integration into current retail frameworks and the creation of intuitive user interfaces. With its potential to give individualised, immersive, and captivating shopping experiences, this technology can completely change the retail industry. Mixed reality stands out as a crucial technology propelling the retail industry's ongoing change, indicating a time when shopping will be more than just a transaction – rather, it will be an immersive exploration of digital innovation (Uwaoma et al., 2023).

MR technology might offer the potential to effectively collaborate designs for prefabrication, site installation, and the organisation and performance of maintenance tasks. Applications of MR technology include "augmented" control interfaces to traditional machinery, remote viewers, or automatic systems, in addition to passive or active rendering and manipulation of online data, such as 3D design prototypes. For instance, a digital prediction of underground utilities recorded in accordance with the real-world correlated structure may be added to an equipment operator's vision from the cab of a traditional backhoe excavator if the operator had an optical see-through head-mounted display (HMD) (Dunston & Wang, 2005).

12.4 APPLICATION OF XR IN MAJOR FASHION MARKETING AREAS

The fashion industry relies heavily on innovation. However, as the retail environment evolves quickly, attracting consumer interest and enhancing the shopping experience remain significant challenges. This is where extended reality comes into play. XR technologies offer a revolutionary approach to streamlining the fashion process, from selecting products online to trying them on and making purchases (Figure 12.2). Extended reality provides engaging and immersive experiences that bridge the gap between online shopping and real-world interaction, enabling customers to make confident and informed choices.

12.4.1 Virtual Fitting

A virtual fitting room is a technology-driven solution that allows consumers to digitally preview clothing, accessories, or makeup using augmented reality or artificial intelligence. This advancement enhances both online and in-store shopping by visually visualising how items look and fit, thereby reducing the need for physical trials. One remarkable creation in this area is the "magic mirror" or "virtual mirror," making dress fitting effortless. The system employs advanced technology, including three-dimensional body scanning, to accurately capture an individual's body shape. Subsequently, it overlays digital garments onto the three-dimensional model in real time, enabling users to virtually "try on" an array of outfits without necessitating any physical alterations. Presented directly on the mirror display, this system frequently

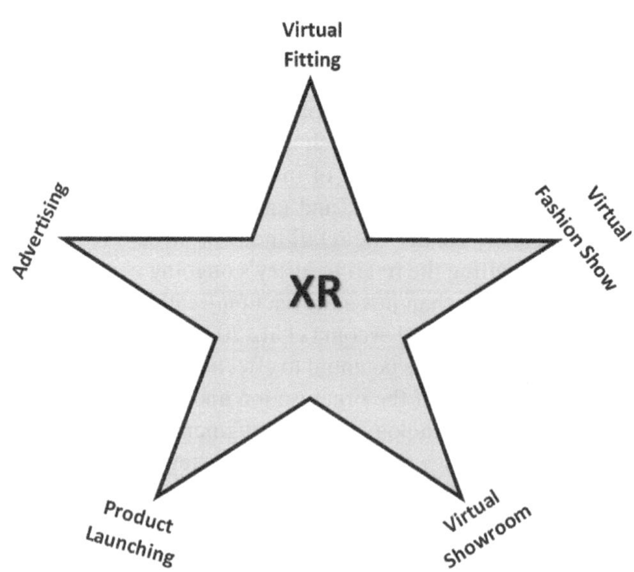

FIGURE 12.2 Areas of fashion where XR can be applied.

incorporates size customisation, colour variations, and style selections, which can be effortlessly managed through a touch interface integrated into the mirror. Smart mirrors designed for dress fitting utilise state-of-the-art technology, such as three-dimensional body scanning, to record a person's body shape precisely. This then projects digital apparel onto the three-dimensional model in real-time, permitting users to virtually "try on" various outfits without requiring physical modification. The mirror's display showcases these options, often featuring size adjustments, colour variations, and style selections, all conveniently modifiable via a touch interface (Idrees et al., 2024).

Customers provide body measurements manually or through body scanning technology. The system captures the user's body shape using 3D scanning technology. The customer selects a garment or item to try on virtually, and the system overlays the selected garment onto the 3D body model in real time. Customers can customise the garment including size, colour, style options, and others. The system adjusts the garment onto the virtual model, and the user "tries on" the garment via the virtual mirror, seeing how it fits and moves. The customer reviews the fit and appearance, can compare it with other garments or styles, and can decide whether to purchase, continue browsing, or try on different items.

12.4.2 Virtual Product Launching

AI and XR have become essential tools in modern marketing. AI empowers businesses to process vast amounts of data, anticipate customer behaviour, and deliver highly personalised content. XR creates immersive experiences that enhance consumer engagement and leave a lasting impression. The extended reality market is expected to exceed $296.9 billion by 2025 and potentially transform various industries, including fashion and entertainment, with AI generating 30 percent of all marketing content (Claudio, 2025). Additionally, 60 percent of marketers already incorporate AI into content creation. Embracing these technologies provides a competitive edge and aligns businesses with a rapidly expanding trend, redefining high-quality advertising (Shi, 2024). When launching a product with XR, users can visualise how the product integrates into their space by superimposing a digital model onto their real-world environment using their smartphone camera. This approach provides a fully immersive experience that allows users to engage with a product virtually, whether by exploring a new car design or taking a digital tour to see its features. It requires to organise an online event where presenters showcase product features enhanced with augmented reality visuals. Virtual product launch creates engaging AR filters for platforms like Instagram or Snapchat, allowing users to try on products virtually or participate in branded challenges.

12.4.3 Virtual Fashion Shows

A virtual fashion show is a digital runway event where brands showcase their collections using extended reality technologies such as VR, AR, and 3D animations. These events eliminate geographical barriers, allowing audiences worldwide to participate remotely. Virtual fashion shows feature a 360-degree video of live fashion show, and participants have reported experiencing a sense presence, feeling as if they were physically at the event location where the fashion show was taking place (Zhang et al., 2023). Virtual fashion shows are open to everyone, unlike traditional fashion events limited to guests. Integrated with gaming or the metaverse, users can participate from anywhere by registering online. In March 2022, *Decentraland's Metaverse Fashion Week* attracted 108,000 attendees, offering virtual brand showcases, shopping, and NFT purchases. By eliminating barriers like invitations and location, virtual fashion shows create a more inclusive and diverse audience (H&M, 2022). The fashion industry, particularly fast fashion, is a major contributor to excessive waste and overproduction. This issue is illustrated by companies like H&M and Burberry, which have been known to dispose of unsold stock. By leveraging 3D models and digital technology, virtual fashion shows can replace the need for physical textiles. This allows designers to reuse and share digital assets, significantly reducing emissions and costs. Participants can engage in these shows through computers or mobile devices. Consequently, virtual fashion shows offer an eco-friendly, low-carbon alternative to traditional fashion events (H&M, 2022). In virtual fashion shows, designers can explore digital fashion, innovative concepts, and AI-generated models, which enhance storytelling and artistic expression. Additionally, audiences can engage in real time through live chats, social media integrations, and direct shopping features, making the event more dynamic and commercially viable.

12.4.4 Virtual Showroom

Virtual showrooms leverage virtual reality to simulate in-person shopping experiences online, particularly for furniture retailers. They focus on creating an interactive interface that allows users to explore, view, and engage with products similarly to how they would in a physical store. Research shows that Virtual Reality Modeling Language (VRML) technology facilitates highly immersive and visually appealing web-based applications. Assessments indicate that 3D VR models improve customer satisfaction compared to conventional 2D catalogues, leading to increasing consumer engagement and enthusiasm about making purchasing. Major e-commerce platforms like Amazon, eBay, and Tmall have incorporated advanced technologies such as 3D modelling, augmented reality, and virtual reality to enhance the shopping experience, particularly for high-return items like clothing. However, the implementation of virtual showrooms comes with challenges. While they may improve the accuracy of the product fit, there is a risk that they could diminish perceived value, which can discourage purchases. (Song et al., 2021). In April 2022, the giant fast-fashion retail H&M introduced its first virtual showroom alongside the launch of the innovation story cherish waste collection. The virtual showroom allowed press and media guests to explore the space, interact with others, and view the collection, which featured high-quality visuals, including photo-realistic 3D-rendered fabrics and garments animated to demonstrate the movement of the innovative materials (Song et al., 2021).

12.4.5 Virtual Advertising

XR in advertising enhances online shopping by allowing customers to virtually try on clothes, cosmetics, and footwear, alleviating size apprehensions and diminishing return rates while augmenting avenues. It offers significant customer insights through analysing preferences and behaviour, enabling personalised marketing. The interactive experience strengthens brand loyalty by promoting engagement and instilling confidence in purchase decisions. VR technology is transforming fashion advertising by offering immersive and interactive experiences that captivate consumers in innovative ways. Fashion brands utilise VR technology to develop virtual showrooms, digital fashion shows, and engaging shopping experiences, strengthening customer interaction and brand loyalty. As VR technology continues to evolve and become more affordable, 360-degree VR content is attracting increasing interest. Since 2015, platforms such as YouTube and Facebook have begun endorsing 360-degree videos, while prominent media entities such as *the New York Times* have integrated VR storytelling into their news reporting. Global investment in VR content and location-based VR was anticipated to increase markedly indicating that VR experiences may soon be broadly available, especially via smartphones. The COVID-19 pandemic significantly increased the consumption of VR content as individuals spent extended periods at home (Goldenberg, 2023).

12.5 CASE STUDIES OF XR IN FASHION

Fashion brands are progressively engaging with the Metaverse to reach Millennials and Gen Z, who are spending more time in digital spaces. Beyond engagement, virtual fashion promotes sustainability by substantially reducing carbon emissions and water usage relative to traditional apparel. We can explore the luxury fashion brand Gucci's journey into the metaverse. Gucci has been a pioneer in Metaverse and Web3, leveraging digital innovation to connect with a global audience. In 2021, it became the first luxury brand to launch an NFT (non-fungible token) and partnered with *Roblox* to create Gucci Garden, an immersive experience that attracted 20 million visitors. In 2022, it launched Super Gucci NFTs, generating $20 million in sales, auctioned an NFT film inspired by its Aria collection, and acquired virtual land in The Sandbox to launch Gucci Vault, a digital retail and gallery. In 2023, it launched Gucci Cosmos Land, a virtual extension of its London exhibition. By integrating NFTs, blockchain, and interactive digital spaces, Gucci continues to redefine luxury fashion in the virtual world (Goldenberg, 2023). The fast-fashion retail brand H&M teamed up with Magic Leap to launch the "Redesign Lab," an interactive experience where customers could design custom prints for existing garments using a headset, providing valuable insight into their preferences. The concept developed through a partnership with Disney, providing VIP customers the chance to design *Star Wars*-themed clothing. Deckers partnered with Rest AR and Unity to digitally scan and render their products, cutting costs by 80 percent and obviating the necessity for physical samples. This approach accelerated production, reduced travel and shipping expenses, and enabled sales representatives to showcase 3D samples in virtual showrooms, enhancing customer insights and decision-making (VOGUE, 2024).

12.6 CONCLUSION

In conclusion, extended reality (XR), which includes virtual reality (VR), augmented reality (AR), and mixed reality (MR), has a significant impact on fashion in areas such as consumer engagement, marketing, sustainability, and retail strategies. This chapter illustrates how XR technologies empower immersive and personalised consumer experience, improving customer engagement through virtual fitting rooms, virtual product launches, and digital fashion shows.

Brands like Gucci and H&M validate XR's commercial viability and transformative potential across luxury and mass markets, while specific case studies demonstrate real-world applications. For instance, Gucci's ventures into digital spaces such as the Metaverse and NFTs also highlight how tech-industry innovation can reinforce its status as a luxury brand. At the same time, H&M's virtual showrooms and digital fashion projects demonstrate XR's potential to increase accessibility and sustainability by reducing physical resources.

However, the global adaptation of XR faces several challenges, including the need for standardisation and technological maturity, exorbitant costs, consumer skepticism, and the transition of business models, including pricing and payment systems. Further technology, targeted consumer education, and partnerships across industries to create best practices need attention to overcome these barriers.

Future research avenues should focus on consumer acceptance of XR technologies, sustainability impact assessments, and the development of strategic roadmaps or frameworks for seamlessly embedding XR within pre-existing retail models. Ultimately, merely adopting and implementing XR innovations within the fashion industry is insufficient; it requires strategic investments driven by analytics that account for evolving consumer behaviours to remain relevant in the future.

REFERENCES

Amed, I., Berg, A., Balchandani, A., Altable, C. S., Harreis, H., Petersens, S. A., & Crump, H. (2022). *The State of Fashion Technology*. Business of Fashion, McKinsey & Company.

Chen, H., Li, H., & Pirkkalainen, H. (2024). How extended reality influences e-commerce consumers: A literature review. *Electronic Commerce Research and Applications*, *65*. https://doi.org/10.1016/j.elerap.2024.101404

Claudio. (2025). *How AI and XR Are Redefining Marketing for Companies*. https://yordstudio.com/how-ai-and-xr-are-redefening-marketing-for-companies/

Dunston, P. S., & Wang, X. (2005). Mixed reality-based visualization interfaces for architecture, engineering, and construction industry. *Journal of Construction Engineering and Management*, *131*(12), 1301–1309. https://doi.org/10.1061/(asce)0733-9364(2005)131:12(1301)

Dunston, P. S., Wang, X., Billinghurst, M., & Hampson, B. (2017). Mixed reality benefits for design perception. *Proceedings of the 19th International Symposium on Automation and Robotics in Construction (ISARC)*, September 2002. https://doi.org/10.22260/isarc2002/0030

Fast-Berglund, Å., Gong, L., & Li, D. (2018). Testing and validating extended reality (xR) technologies in manufacturing. *Procedia Manufacturing*, *25*, 31–38.

Gabriel, A., Ajriya, A. D., Fahmi, C. Z. N., & Handayani, P. W. (2023). The influence of augmented reality on E-commerce: A case study on fashion and beauty products. *Cogent Business and Management*, *10*(2). https://doi.org/10.1080/23311975.2023.2208716

Goldenberg, B. (2023). *Gucci XR/Metaverse Case Study*. https://ismguide.com/gucci-xr-metaverse-case-study/

H&M. (2022). *A Virtual Reality at H&M Group – H&M Group*. https://hmgroup.com/our-stories/a-virtual-reality-at-hm-group/

Hackl, C., & Wolfe, S. G. (2017). *Marketing New Realities: An Introduction to Virtual Reality & Augmented Reality Marketing, Branding, & Communications*. Meraki Press. ISBN-13: 978-0996510677.

Hillmann, C. (2021). *UX for XR: User Experience Design and Strategies for Immersive Technologies*. Springer.

Hughes, C. E., Stapleton, C. B., Hughes, D. E., & Smith, E. M. (2005). Mixed reality in education, entertainment, and training. *IEEE Computer Graphics and Applications*, *25*(6), 24–30. https://doi.org/10.1109/MCG.2005.139

Idrees, S., Gill, S., & Vignali, G. (2024). Mobile 3D body scanning applications: A review of contact-free AI body measuring solutions for apparel. *The Journal of The Textile Institute*, *115*(7), 1161–1172.

Jung, J., Yu, J., Seo, Y., & Ko, E. (2021). Consumer experiences of virtual reality: Insights from VR luxury brand fashion shows. *Journal of Business Research*, *130*(October), 517–524. https://doi.org/10.1016/j.jbusres.2019.10.038

Jung, T., Bae, S., Moorhouse, N., & Kwon, O. (2024). The effects of experience-technology fit (ETF) on consumption behavior: Extended reality (XR) visitor experience. *Information Technology and People*, *37*(5), 2006–2034. https://doi.org/10.1108/ITP-01-2023-0100

Kim, S. J. (2023). Virtual fashion experiences in virtual reality fashion show spaces. *Frontiers in Psychology*, *14*(November), 1–13. https://doi.org/10.3389/fpsyg.2023.1276856

Koontz, M. L., & Gibson, I. E. (2002). Mixed reality merchandising: Bricks, clicks – and mix. *Journal of Fashion Marketing and Management*, *6*(4), 381–395. https://doi.org/10.1108/13612020210448664

Li, N., Asce, M., Du, J., González, V. A., Chen, J., & Asce, S. M. (2022). *Methodology for Extended Reality-Enabled Experimental Research in Construction Engineering and Management*. https://doi.org/10.1061/(ASCE)CO.1943

Logaldo, M. (2016). Augmented bodies: Functional and rhetorical uses of augmented reality in fashion. *Pólemos*, *10*(1), 125–141. https://doi.org/10.1515/pol-2016-0007

Morotti, E., Stacchio, L., Donatiello, L., Roccetti, M., Tarabelli, J., & Marfia, G. (2022). Exploiting fashion x-commerce through the empowerment of voice in the fashion virtual reality arena: Integrating voice assistant and virtual reality technologies for fashion communication. *Virtual Reality*, *26*(3), 871–884. https://doi.org/10.1007/s10055-021-00602-6

Pan, Z., Cheok, A. D., Yang, H., Zhu, J., & Shi, J. (2006). Virtual reality and mixed reality for virtual learning environments. *Computers and Graphics (Pergamon)*, *30*(1), 20–28. https://doi.org/10.1016/j.cag.2005.10.004

Park, M., Im, H., & Kim, D. Y. (2018). Feasibility and user experience of virtual reality fashion stores. *Fashion and Textiles*, *5*(1). https://doi.org/10.1186/s40691-018-0149-x

Quint, F., Sebastian, K., & Gorecky, D. (2015). A mixed-reality learning environment. *Procedia Computer Science*, *75*(Vare), 43–48. https://doi.org/10.1016/j.procs.2015.12.199

Reality, A. (2011). Handbook of augmented reality. In *Handbook of Augmented Reality*. https://doi.org/10.1007/978-1-4614-0064-6

Reddy, S. R. (2024). *Virtual Vogue: Unravelling the Impact of Augmented Reality on Consumer Motivations in Online Fashion* (Doctoral dissertation, Auckland University of Technology).

Sabatini, N., Sádaba, T., Tosi, A., Neri, V. & Cantoni, L. (Eds) (2023). *Fashion Communication in the Digital Age*. Proceedings of the FACTUM 23 Conference, Pisa, Italy, 2023. Cham: Springer. https://doi.org/10.1007/978-3-031-38541-4

Sarakatsanos, O., Papazoglou-Chalikias, A., Boikou, M., Chatzilari, E., Jauk, M., Hafliger, U., Nikolopoulos, S., & Kompatsiaris, I. (2024). VR designer: Enhancing fashion showcases through immersive virtual garment fitting. *Virtual Reality*, *28*(2), 1–16. https://doi.org/10.1007/s10055-024-00945-w

Shi, Y. (2024). Application research of virtual fashion shows. *2nd International Conference on Humanities and Arts, Design and Culture*, *10*, 119–124. https://doi.org/10.62051/mbrcyt03

Silvestri, B. (2022). How virtual and augmented reality are reshaping the fashion industry during the covid-19 pandemic. In *Extended Reality Usage during COVID 19 Pandemic* (pp. 39–54). Springer.

Song, H., Kim, J., Nguyen, T. P. H., Lee, K. M., & Park, N. (2021). Virtual reality advertising with brand experiences: the effects of media devices, virtual representation of the self, and self-presence. *International Journal of Advertising*, *40*(7), 1096–1114.

Spadoni, E., Fiocca, A., Zoni, G., Infante, L. M. U., Cerutti, L., Maccarrone, P., Carulli, M., & Bordegoni, M. (2024). A virtual reality experience to raise sustainability awareness within the fashion industry. *Proceedings of the Design Society*, *4*, 1447–1456. https://doi.org/10.1017/pds.2024.147

Stanney, K. M., Nye, H., Haddad, S., Hale, K. S., Padron, C. K., & Cohn, J. V. (2021). Extended reality (XR) environments. *Handbook of Human Factors and Ergonomics*, 782–815.

Starkey, S., Alotaibi, S., Striebel, H., Tejeda, J., Francisco, K., & Rudolph, N. (2021). Fashion inspiration and technology: Virtual reality in an experimental apparel design classroom. *International Journal of Fashion Design, Technology and Education*, *14*(1), 12–20. https://doi.org/10.1080/17543266.2020.1844807

Uwaoma, P. U., Eboigbe, E. O., Eyo-Udo, N. L., Ijiga, A. C., Kaggwa, S., & Daraojimba, A. I. (2023). Mixed reality in U.S. retail: A review: Analyzing the immersive shopping experiences, customer engagement, and potential economic implications. *World Journal of Advanced Research and Reviews*, *20*(3), 966–981. https://doi.org/10.30574/wjarr.2023.20.3.2495

VOGUE. (2024). Gucci retains Index innovation leadership, but Hugo Boss and Versace surprise with virtual activations. *Vogue Business*. https://www.voguebusiness.com/story/technology/gucci-retains-index-leadership-but-hugo-boss-and-versace-surprise-with-virtual-activations

Wu, Y. F., & Kim, E. Y. (2022). Users' perceptions of technological features in augmented reality (AR) and virtual reality (VR) in fashion retailing: A qualitative content analysis. *Mobile Information Systems, 2022*.

Zhang, T., Li, G., & Tayi, G. K. (2023). A strategic analysis of virtual showrooms deployment in online retail platforms. *Omega*, *117*, 102824. https://doi.org/https://doi.org/10.1016/j.omega.2022.102824

Part D

Fashion Journey in Metaverse

13 Metaverse Meets Fashion
The Fusion of Creativity and Technology

Chenming Lin, Yuyuan Shi, Guobin Xia, Gary Pogue, Linda Lightley, and Sheng-feng Qin

13.1 INTRODUCTION

The goal of digitizing the fashion industry was to streamline the design, production, and commercial aspects of physical products in the real world, whilst also advancing sustainability through diverse digital tools. However, with the recent advent of the metaverse, a parallel virtual reality world, a new frontier in digital fashion has been revealed. This advancement has been significantly aided by the introduction of three-dimensional computer-aided design (3D CAD) platforms in the clothing industry, which are typically categorized into two types. One type allows designers to create garments directly in a three-dimensional (3D) virtual environment, tailored to personal priorities – such as TPC Parametric Pattern Generator (TPC, HK). The other type simulates 3D design from imported 2D information, for instance, Accumark VstistcherTM (Gerber) and Modaris 3D FIT (Lectra) (Sayem et al., 2010). Additionally, reverse engineering software systems, such as Geomegic package (3D Systems, Inc.) and the DesignConcept software (Lectra), have been introduced in the clothing field (Sayem et al., 2010; Sayem, 2017; Shi et al., 2023) to facilitate the development of innovative technologies. The 3D CAD system is also treated as a tool to reduce material waste and support zero waste in garment design and manufacturing process (Mcquillan, 2020). Here, zero waste refers to minimizing or preventing waste during the cutting and sewing process (Townsend & Mills, 2013; Carrico et al., 2022). CLO3D software (CLO Virtual Fashion Inc.) is the primary tool used in the design process. The traditional apparel-making process is a trial-and-error-and-correct process (James et al., 2016). This process is repeated until achieving the proposed design. CLO3D facilitates the conversion between 2D pattern pieces and 3D garment visualizations, allowing designers to either drape flat patterns onto a 3D avatar or flatten a 3D garment surface back into 2D pattern pieces. This helps to shorten the manufacturing process and may reduce the trial material waste (Mcquillan, 2020).

The term *metaverse* describes an expansive future realm within a digital setting, originating from the fusion of "meta" (beyond) and "universe" (virtual world). Reflective of its name, the metaverse can be divided into two key areas: the first is

DOI: 10.1201/9781003594482-17

a virtually augmented reality where individuals, through avatars, engage in human interaction and socio-cultural exchanges that may outline a functioning economy. The second area involves technological integrations that blend various contemporary technologies and social media ideas to forge a virtual space that surpasses the physical world (Dionisio et al., 2013). In the area of fashion industry, it is forecasted that the metaverse will comprehensively transform the entire fashion design process. Whereas the contemporary internet facilitates the asynchronous transmission of 2D information in fashion design, the metaverse offers a persistent, shared virtual space accessible through a variety of devices, platforms, and media touchpoints, thereby enabling synchronous interactions between fashion designers, manufacturers, consumers, and brands (Yoo et al., 2023). The metaverse is frequently discussed in conjunction with Web 3.0, a decentralized iteration of the internet that prioritizes community and user autonomy (Cheng, 2023). It is critical to acknowledge that the metaverse is in its nascent stages, with the opportunity to develop immersive virtual worlds, retail environments, and societies just beginning to materialize. Given that the metaverse is projected to catalyze the most significant industrial revolution witnessed to date, there is a pressing demand for the fashion design process to carve out and delineate their niches within this domain. Some studies also demonstrate that a well-designed virtual environment can positively influence people's cognitive performance. This suggests that the effective use of metaverse environments in fashion design could enhance designers' lateral thinking abilities, potentially leading to better fashion designs (Xia et al., 2021, 2023). However, additional research is imperative to elucidate the potential uses and capabilities of the metaverse as these novel technologies mature. This research aims to explore how fashion design can effectively navigate this novel space by reviewing current practices to identify new opportunities for design innovation.

13.2 METHODS

The integration of virtual reality (VR) and 3D modelling has revolutionized the fashion industry, offering new ways to create, present, and experience garments. The rise of digital tools and immersive technologies has not only transformed traditional garment production workflows but also provided new opportunities for interactive and engaging consumer experiences. Virtual fashion showcases and digital design pipelines are becoming increasingly prevalent, allowing designers to create garments in a fully virtual space and present them in novel, engaging formats (Zhang et al., 2023). This method offers distinct advantages, such as reducing material waste, accelerating the design cycle, and providing an immersive platform for consumers to interact with fashion (Särmäkari, 2023; Adekunle, 2024).

The current methodology outlines a comprehensive workflow for the development of an immersive virtual fashion showcase. The pipeline utilizes a range of tools and software, including CLO3D for garment modelling and simulation, Adobe Photoshop for texturing, Autodesk Maya for animation refinement, and Unity3D as the platform for integrating and showcasing the final results within a virtual showroom. Each of these tools serves a critical role in the design process, from garment creation to interactive presentation, ensuring the resulting product is both visually appealing and functionally engaging.

The approach described here combines advanced software applications to support a fully digital workflow. By using CLO3D's capabilities for real-time garment simulation and Maya's animation refinement tools, designers can bring virtual garments to life, showcasing them through interactive and immersive experiences. Unity3D, with its VR development support, provides the final environment where these designs can be experienced by audiences, bridging the gap between fashion design and consumer engagement in the digital realm. This methodology not only highlights the power of integrated digital tools but also demonstrates how immersive technologies can enhance creative industries, offering new ways to conceptualize and experience fashion.

13.2.1 Tools and Software

In the design process, we use 3D modelling software to generate a character model, followed by designing the required garments. Specialized software is then employed to refine the runway animation, allowing the character to effectively showcase the garments through animation. Finally, all components are integrated using a video game engine to add interactivity within the virtual showroom. The software used in this sequence includes CLO3D, Photoshop, Maya, and Unity.

CLO3D is a sophisticated software application designed for the fashion industry, enabling the creation, simulation, and visualization of garments in a 3D environment. CLO3D allows designers to create virtual clothing that can be adjusted and visualized in real time, facilitating a more efficient design cycle compared to traditional methods. The software's utility extends beyond mere design; it also plays a crucial role in the evaluation of garment fit and performance. CLO3D incorporates advanced physical properties of fabrics, such as stretch, shear, and bending characteristics, which allow for realistic simulations of how garments will behave in real life (Huang & Huang, 2022). This functionality is essential for designers aiming to create clothing that not only looks good but also fits well and performs effectively under various conditions. Moreover, CLO3D supports the development of virtual clothing display platforms, enabling designers to showcase their creations online, thus enhancing customer engagement and feedback (Wang & Liu, 2020).

Adobe Photoshop is a powerful image editing software that has become a staple in various fields, including graphic design, photography, and digital art. Since its initial release in 1988, Photoshop has evolved into a comprehensive tool that offers a wide range of features for manipulating images and creating complex graphics. Its advanced capabilities allow users to perform tasks such as photo retouching, colour correction, compositing, and the creation of digital artwork (Nomura et al., 2018).

Autodesk Maya is a comprehensive 3D modelling, animation, simulation, and rendering software. The Graph Editor (Autodesk, 2024) in Autodesk Maya is a powerful tool that significantly enhances the animation workflow by allowing animators to visualize and manipulate animation keyframes in a more intuitive and efficient manner. This feature is essential for refining animations, as it provides a graphical representation of the motion curves associated with animated attributes, enabling precise control over timing and interpolation between keyframes.

Unity3D is a professional cross-platform game engine and development tool that excels in various aspects such as 3D content design, virtual reality, and 3D visualization (Wang, 2023).

13.2.2 Process

In this section, we provide a detailed overview of the process involved in creating the virtual fashion showcase; the design process is illustrated in Figure 13.1. First, we needed to add a virtual avatar to our project. CLO3D provides free avatars, including male figures, child figures, and female figures, all initially presented in an A-pose stance (with both arms extended downward at a 45-degree angle), which meets the requirements of most projects (Figure 13.2a). The A-pose can facilitate the creation of a more natural range of motion for the character's arms, as it allows for a more relaxed positioning compared to the T-pose (with both arms extended horizontally to the sides) (Zhan et al., 2024; Li et al., 2021). It also offers a wide range of customization modules, allowing adjustments to parameters such as "Neck base", "Bust", "Bicep", "Waist", "High hip", "Low hip", and "Thigh" to customise the avatar to the needs of the project (Figure 13.2b). In this case, the default female avatar was selected to support the subsequent garment presentation and animation.

Next, we began designing the garment for the character. Creating virtual clothing requires an in-depth study of the garment's pattern. Users can purchase and download similar garment patterns from CLO3D's official asset store (connect.clo-set.com) and then modify them based on the specific needs of their projects. The official account of CLO3D, known as "CONNECT Official," offers many basic garment patterns that users can download for free. These pattern collections are provided under a free commercial license by the Korea Federation of Textile Industries (KOFOTI). However, instead of downloading garment patterns from CLO3D's official asset

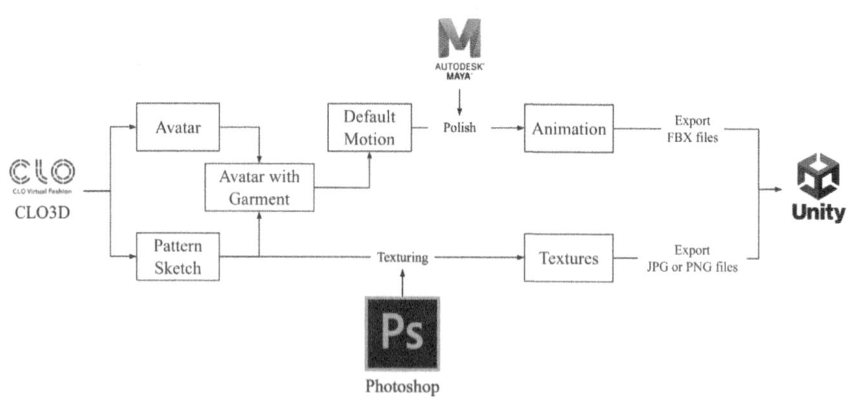

FIGURE 13.1 CLO3D to Unity design process.

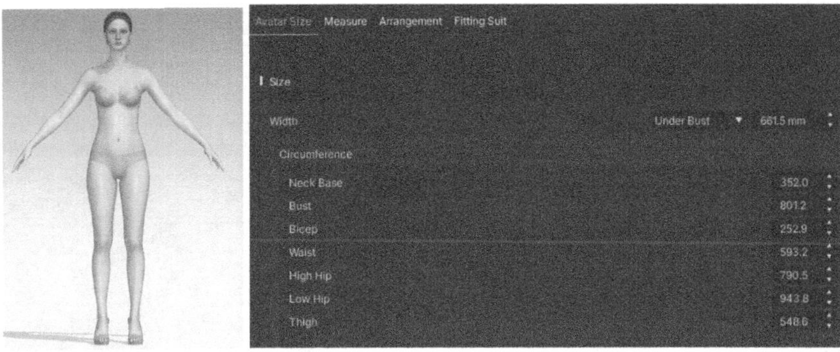

a b

FIGURE 13.2 (a) Free avatar in CLO3D, and users can use (b) Avatar Editor for customization.

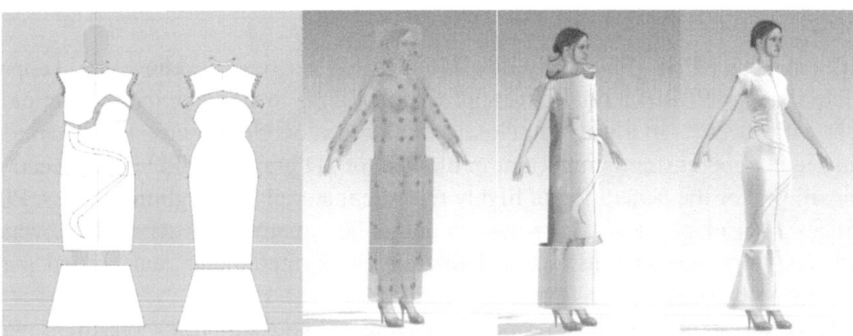

FIGURE 13.3 (a) 2D garment block construction in 2D pattern window, (b) Arrangement feature, (c) 2D pattern in 3D perspective, (d) Clothing simulation on the avatar.

store, this project used an original design created by a Northumbria Fashion student. Specifically, the tight-fitting long dress in this case was from the graduation project "Seen by Dismissed" by Lucy Hall. This approach ensured that the garment design was unique and tailored to the specific needs of this project.

Firstly, we created a 2D garment pattern in CLO3D, as shown in Figure 13.3a. The next step is to wrap the designed 2D garment block onto the mannequin, effectively "dressing" the avatar. Using the Arrangement feature of the Avatar Editor (Figure 13.3b), we placed the fabric pieces in appropriate positions (Figure 13.3c). Finally, by activating the fabric simulation, CLO3D automatically wraps the fabric onto the character, resulting in the dressed appearance shown in Figure 13.3d.

For the dress material and textures, we used 2D design software (e.g., Adobe Photoshop) to create textures for the pattern. We developed three different textures with varying details and fabrics (Figure 13.4a). These textures were then applied to the

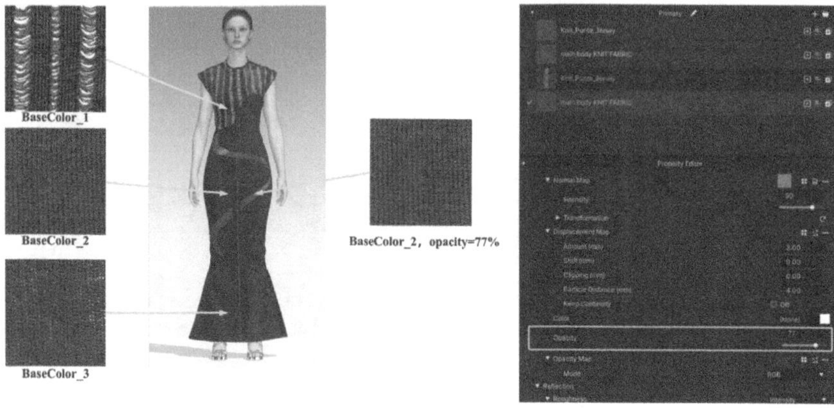

FIGURE 13.4 (a) Assign different color textures to various fabrics, and (b) parametric adjustment of materials.

garment using a PBR (Physically Based Rendering) material workflow to add appropriate shaders. PBR is a rendering approach that aims to simulate the interaction of light with surfaces in a way that closely mimics real-world physics. This technique has become increasingly important in the field of 3D graphics and texture creation, as it allows for the generation of highly realistic materials and lighting effects. PBR utilizes a set of physical parameters to define how materials react to light, which includes properties such as base colour, roughness, metallicity, and normal maps (McDermott, 2018).

Thus, we needed to adjust properties such as colour, bumpiness, opacity, roughness, and metallicity for our garment material. These adjustments were categorized into texture adjustments and parameter adjustments based on whether the property required global consistency. The approach for these adjustments varies depending on the project, as illustrated in this case.

Texture Adjustments: These involve using texture maps to define the visual characteristics of the material. For example, the base colour map is used to determine the primary colour of the garment surface, allowing for colour variations on the fabric. The normal map is used to simulate subtle surface bumps, giving a smooth surface the appearance of greater depth and detail, thus achieving a knit texture effect for the garment.

Parameter Adjustments: These adjustments involve changing the material's physical property values. For instance, the roughness parameter controls the smoothness of the surface, affecting how light is reflected. A lower roughness value results in clearer reflections, while a higher value gives a more diffuse, scattered reflection. Since our garment represents a brand-new piece, the roughness across different areas of the fabric remains consistent, allowing us to adjust the roughness uniformly via parameter values. The metallicity parameter defines the metallic characteristics of the material, influencing light reflection and absorption properties, which distinguish metallic and non-metallic materials visually. Since there are no metallic elements in our garment, we uniformly reduced this parameter. Lastly, the opacity parameter, adjustable from 0 to 100 (with 0 indicating complete transparency and 100 indicating complete opacity), was used to add different textures to specific areas of the fabric. For instance, the curved fabric in the middle of the dress was set to an opacity value of 77 (see Figure 13.4b) to give it a semi-transparent effect, simulating the appearance of a lighter, gauze-like fabric.

Once the mannequin and garment are ready, we use the built-in motion library in CLO3D to add a walking animation to the character. If further polishing of the default walking animation is desired, the animation data can be exported and imported into Maya for refinement using the Graph Editor. One of the primary functions of the Graph Editor is to display the animation curves associated with various attributes, such as position, rotation, and scale. Each keyframe is represented as a point on the curve, and the shape of the curve dictates how the animation progresses between these keyframes. By selecting and adjusting these curves, animators can fine-tune the timing and easing of movements, allowing for greater control over the animation's flow (Lazor et al., 2019). For instance, if an animator wants to create a more dynamic action, they can adjust the tangents of the curve to create a faster acceleration or deceleration, resulting in a more lifelike movement.

The final process is bringing 3D models from MAYA and CLO3D into Unity3D for VR purposes, which involves exporting the models from the original software and importing them into Unity3D for further development. Unity3D's support for various 3D modelling software, its ease of use for model modification, and its application in VR development make it a suitable platform for creating immersive VR experiences.

To establish a structured pipeline for the integration of 3D models into Unity3D for VR applications, the presence of a systematic workflow is of utmost importance. To incorporate 3D models into Unity3D for the purpose of developing games or applications, it is imperative to adhere to particular procedures, sourced from the official documentation of Unity3D.

Unity3D is a versatile platform that facilitates the import of 3D models from various sources, including popular 3D modelling software like MAYA, 3DS Max, and Blender. Imported assets can originate from external files such as 3D models, audio files, or images. Figure 13.5 illustrates the Unity3D pipeline workflow, from importing assets through to loading and distributing the final application to users. The process begins with gathering 3D models created in MAYA and CLO3D, ensuring that these models are optimized and prepared for export to maintain quality during integration. Utilizing the export functionalities within MAYA and CLO3D, the 3D models should be saved in formats compatible with Unity3D, such as FBX or OBJ.

When importing 3D models into Unity3D, it is crucial to recognize the differences between the FBX and OBJ file formats. The FBX (Filmbox) format, developed by Autodesk, is known for its ability to store not only the 3D geometry of the model but also animation data, textures, materials, and other relevant information within a single file. This comprehensive nature of FBX files makes them suitable for complex 3D models requiring animations and detailed textures. Conversely, the OBJ file format is more straightforward, primarily focusing on storing the 3D geometry of the model. OBJ files consist of separate files for the 3D geometrical data, texture

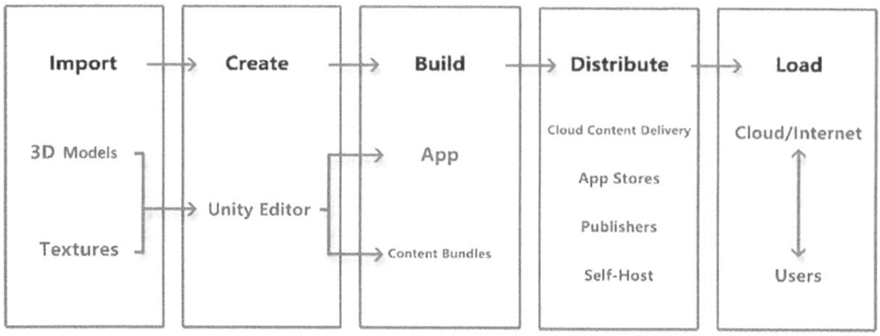

FIGURE 13.5 Asset workflow from Unity official document.

data (such as JPEG or PNG), and material information. Unlike FBX files, OBJ files do not inherently support animations or complex material properties. However, they are widely supported by various 3D modelling software and are commonly used for static 3D models that do not require animations or advanced features.

Once the assets are imported, utilise the Unity Editor to develop and enhance the content as needed for the virtual reality (VR) application. Unity3D provides tools for real-time adjustments, enabling customization and arrangement of models to suit the VR environment. Progress with the development by employing supported VR toolkits, and integrating the imported 3D models into the virtual environment. Unity3D's compatibility with various VR devices, including headsets such as HTC Vive and Oculus Rift, allows developers to target a wide range of platforms for their VR applications (Schlueter et al., 2017). Conduct thorough testing to ensure that the integrated 3D models function correctly within the virtual environment, optimizing performance based on feedback. Once the VR application is finalized and tested, prepare it for deployment on the desired VR platform, ensuring compatibility and functionality across different devices. Updates may be required at runtime based on user behaviour and how content has been grouped and packaged.

By adhering to this structured pipeline, users can effectively bring 3D models from MAYA and CLO3D into Unity3D for VR applications, leveraging the strengths of each software and ensuring a seamless integration process. Combining Unity3D's robust features, compatibility with various devices, and extensive library of assets make it a powerful tool for creating compelling VR experiences.

13.3 CONCLUSIONS

The integration of fashion design within the metaverse signifies a profound shift in the industry, leveraging advanced digital tools to create innovative and sustainable garments. This study demonstrates how three-dimensional computer-aided design (3D CAD) platforms, such as CLO3D, facilitate the development of zero-waste designs by allowing precise pattern-making and virtual garment simulation. These tools streamline traditional trial-and-error processes, significantly reducing material waste and enhancing design precision.

The advent of the metaverse introduces a new realm where designers, manufacturers, and consumers can interact synchronously in a persistent, shared virtual space. This environment fosters creativity and collaboration while enhancing the efficiency of the fashion design process. The ability to create and test virtual garments in a 3D space offers unprecedented flexibility and innovation, pushing the boundaries of what is possible in fashion design. Furthermore, the structured pipeline for integrating 3D models from platforms like MAYA and CLO3D into Unity3D for VR applications illustrates the potential for creating immersive fashion experiences. By adhering to a systematic workflow, designers can ensure a seamless transition from digital models to VR environments, enhancing the overall quality and engagement of virtual fashion showcases.

Despite these advancements, challenges such as digital accessibility, data security, and the environmental impact of maintaining virtual environments need to be addressed. As the metaverse and related technologies continue to evolve, ongoing research and development will be crucial in addressing these issues and fully realizing the potential of this digital frontier. In conclusion, the metaverse offers a transformative platform for the fashion industry, merging creativity with sustainability. By embracing these digital innovations, fashion designers can pioneer new methods of garment creation that are both cutting-edge and environmentally conscious. The future of fashion design lies in the seamless integration of digital tools and virtual environments, promising a more sustainable and innovative industry.

13.4 ACKNOWLEDGEMENTS

The authors would like to express their gratitude to Lucy Hall for her exceptional graduation project, “Seen by Dismissed,” which served as an invaluable example in illustrating the design process in CLO3D for this book chapter. Lucy’s creativity and dedication have greatly enriched the content and provided a practical lens through which readers can understand the application of digital fashion design tools. We are deeply appreciative of her contribution to this work.

REFERENCES

Adekunle, A. (2024). Application of artificial intelligence and digital technologies in fashion design and innovation in Nigeria. *International Journal of Fashion and Design*, 3(1), 37–48.

Autodesk. (2024). *Maya 2025 Help*. https://help.autodesk.com/view/MAYAUL/2025/ENU/

Carrico, M., Dragoo, S.L., McKinney, E., Stannard, C., Moretz, C., & Rougeaux-Burnes, A. (2022). An inquiry into gradable zero-waste apparel design. *Sustainability (Switzerland)*, 14(1).

Cheng, S. (2023). *Metaverse: Concept, Content and Context*. Springer Nature Switzerland AG. https://doi.org/10.1007/978-3-031-24359-2

Dionisio, J.D.N., Burns, W.G. III, & Gilbert, R. (2013). 3D virtual worlds and the metaverse: Current status and future possibilities. *ACM Computing Surveys*, 45(3).

Huang, S., & Huang, L. (2022). CLO3D-based 3D virtual fitting technology of down jacket and simulation research on dynamic effect of cloth. *Wireless Communications and Mobile Computing*, 2022(1), 5835026.

James, A.M., Roberts, B.M., & Kuznia, A. (2016). Transforming the sequential process of fashion production: Where zero-waste pattern cutting takes the lead in creative design. *International Journal of Fashion Design, Technology and Education*, 9(2), 142–152.

Lazor, M., Gajić, D.B., Dragan, D., & Duta, A. (2019). Automation of the avatar animation process in FBX file format. *FME Transactions*, 47(2).

Li, P., Aberman, K., Hanocka, R., Liu, L., Sorkine-Hornung, O., & Chen, B. (2021). Learning skeletal articulations with neural blend shapes. *ACM Transactions on Graphics (TOG)*, 40(4), 1–15.

McDermott, W. (2018). *The PBR Guide*. Allegorithmic.

Mcquillan, H. (2020). *Zero Waste Systems Thinking: Multimorphic Textile Forms*. University of Borås.

Nomura, S., Freitas, K.M.S., Silva, P.P.C.D., Valarelli, F.P., Cançado, R.H., Freitas, M.R.D., . . . Oliveira, R.C.G.D. (2018). Evaluation of the attractiveness of different gingival zeniths in smile esthetics. *Dental Press Journal of Orthodontics*, 23(5), 47–57.

Särmäkari, N. (2023). Digital 3D fashion designers: Cases of atacac and the fabricant. *Fashion Theory*, 27(1), 85–114.

Sayem, A.S.M. (2017). Objective analysis of the drape behaviour of virtual shirt, part 2: Technical parameters and findings. *International Journal of Fashion Design, Technology and Education*, 10(2), 180–189.

Sayem, A.S.M., Kennon, R., & Clarke, N. (2010). 3D CAD systems for the clothing industry. *International Journal of Fashion Design, Technology and Education*, 3(2), 45–53.

Schlueter, J., Baiotto, H., Hoover, M., Kalivarapu, V., Evans, G., & Winer, E. (2017). Best practices for cross-platform virtual reality development. In *Degraded Environments: Sensing, Processing, and Display 2017*, Proc. SPIE 10197, 1019709. https://doi.org/10.1117/12.2262718

Shi, Y., Taylor, L.W., Cheung, V., & Sayem, A.S.M. (2023). *Scan2Weave: Connecting Digital Anthropometry with 3D Weaving Technology*. CRC Press, pp.83–96.

Townsend, K., & Mills, F. (2013). Mastering zero: How the pursuit of less waste leads to more creative pattern cutting. *International Journal of Fashion Design, Technology and Education*, 6(2), 104–111.

Wang, Y.X., & Liu, Z.D. (2020). Virtual clothing display platform based on CLO3D and evaluation of fit. *Journal of Fiber Bioengineering and Informatics*, 13(1), 37–49.

Xia, G., Henry, P., Chen, Y., Queiroz, F., Westland, S., & Cheng, Q. (2023). The effects of colour attributes on cognitive performance and intellectual abilities in immersive virtual environments. *Computers in Human Behavior*, 148(July), 107853.

Xia, G., Henry, P., Queiroz, F., & Westland, S. (2021). Effects of coloured lighting in the real world environment and virtual reality. *Journal of the International Colour Association*, 27(June), 9–25.

Yoo, K., Welden, R., Hewett, K., & Haenlein, M. (2023). The merchants of meta: A research agenda to understand the future of retailing in the metaverse. *Journal of Retailing*, 99(2), 173–192.

Zhan, X., Fu, R., & Ritchie, D. (2024, April). Character mixer: Rig-aware interpolation of 3D characters. *Computer Graphics Forum*, 43(2), p. e15047.

Zhang, Y., Liu, C., & Lyu, Y. (2023). Examining consumers' perceptions of and attitudes toward digital fashion in general and purchase intention of luxury brands' digital fashion specifically. *Journal of Theoretical and Applied Electronic Commerce Research*, 18(4), 1971–1989.

14 Exploring the Affective Quality of Virtual Fashion Design in the Metaverse Era

Eun Kyoung Yang, Jeong Min Kho, and Ruobing Li

14.1 INTRODUCTION: BACKGROUND AND RESEARCH PURPOSE

The Metaverse era merges technology and fashion, initiating a digital transformation that gives rise to virtual fashion, where creativity and technology intersect. This convergence has redefined how fashion is conceptualized, created, and experienced in digital environments, positioning virtual fashion as a key medium for self-expression and user engagement (Santos et al. 2024). As the Metaverse continues to evolve, virtual fashion not only facilitates creativity but also fosters emotional resonance through affective qualities of core design elements. These elements – colors, shapes, and textures – play a crucial role in shaping user perceptions, enhancing self-expression, and influencing psychological responses such as aesthetic satisfaction and a sense of belonging (Khelladi et al. 2023). While prior research has largely focused on the functional and strategic aspects of virtual fashion, research on the emotional dimensions embedded in virtual fashion design remains limited. Addressing this gap, this study explored how core design elements contribute to users' emotional engagement with virtual fashion. Through a mixed-method approach – including user surveys, emotional response analysis, and case studies of virtual fashion from leading brands – this research identified key affective qualities linked to core design elements of virtual fashion. By examining emotional resonance from a design perspective, the study offered theoretical insights and strategic applications for enhancing virtual fashion experiences in the Metaverse era.

14.2 RELATED WORKS

Virtual fashion design is reshaping the way fashion is created and experienced, emerging as a key force in the fashion industry's digital transformation. Both established brands and innovative startups are embracing this evolution, introducing virtual garments and accessories that redefine design possibilities. Notable examples include Gucci's AR-compatible Virtual 25 sneaker in 2021 and Burberry's *Roblox* Lola handbag in 2022, both featuring digital materials to expand fashion's creative

DOI: 10.1201/9781003594482-18

boundaries (Loureiro 2023). Companies such as The Fabricant, RTFKT, and Dress X further push these boundaries by integrating virtual fashion into gaming, social media, and NFT-based ecosystems. The Fabricant provides a fully virtual fashion supply chain, allowing users to create digital identities through personalized garments, while RTFKT has developed a digital ecosystem involving NFTs and metaverse games, and Dress X offers virtual try-ons that support self-expression and sustainability (Cascianni et al. 2022; Hollensen et al. 2023). The Metaverse has amplified virtual fashion's impact, incorporating VR, AR, and blockchain technologies to enhance self-expression and emotional engagement within digital spaces (Kye et al. 2021). However, while studies have explored the psychological motivations behind virtual fashion consumption (Venturini and Columbano 2023) and its role in brand strategy within the Metaverse (Zaki et al. 2024), little research has focused on how specific design elements evoke emotional engagement. Addressing this gap is critical, as virtual fashion's affective qualities – its ability to evoke emotional responses through colors, shapes, and textures – play a fundamental role in users' emotional engagement (Norman 2004). Understanding these emotional dimensions not only enhances design strategies but also enable brands to create digital fashions that foster deeper emotional connections with users (Kim and Kim 2015), as affective qualities engage the senses and shape consumer perceptions.

This connection is shaped by key design elements such as color, shape, and texture, which together define a product's sensory and emotional appeal (Kodzoman et al. 2023). Unlike conventional aesthetics, affective design emphasizes emotional interaction with products, fostering long-term engagement (Norman 2004; Stauss 2023). Previous studies have developed frameworks for measuring emotional responses in fashion. An and Lee (2015) introduced 25 emotional adjective pairs (e.g., 'soft-hard,' 'complex-simple,' 'feminine-mannish,' 'modern-classic,' 'warm-cool') to assess global consumers' preferences, while Jin and Choi (2023) employed perceptual engineering to map design attributes (lines, shapes, textures, colors, and patterns) to emotional preferences of Gen Z consumers in China. These findings highlight the importance of cultural context and personalized needs in emotional design, highlighting the need for tailored approaches in various user groups.

In virtual fashion, affective qualities are vital for self-expression and emotional resonance (Yang et al., 2025). Norman's emotional design theory (2004) explained how design elements shape user experience, a concept now redefined by advances in digital technologies. Virtual imagery has been particularly recognized as a potent tool for expressing emotional elements, creating intimate interactions between fashion products and consumers. Kim and Kim (2022) introduced deep learning to enhance developer-user emotional communication, while Ma et al. (2022) developed a data-driven system that allows consumers to transform emotional descriptors into tangible fashion design elements, which meet personalized fashion needs. Despite these advancements, research on affective design in virtual fashion is limited. The Metaverse merges fashion with gaming, social media, and immersive tech, requiring strategic design approaches to embed emotional qualities in virtual fashion. By prioritizing affective design, virtual fashion can enhance self-expression and emotional connection, shaping its role in the digital fashion economy.

14.3 METHODOLOGY

The study examined the affective qualities of color, shape, and texture in virtual fashion design by analyzing their emotional influence on emotional engagement. The research followed a structured four-phase approach: (1) collecting 3,121 virtual fashion designs from five prominent brands (RTFKT, Republiqe, the Fabricant, Tribute, and Dress X), (2) identifying and refining 8,031 emotional adjectives through expert evaluation, (3) validating a final set of 61 emotional adjective pairs by assessing their relevance, and (4) mapping their associations with key design elements to assess their emotional resonance.

A dataset of 3,121 virtual fashion designs from five leading brands was compiled, covering five seasons from January 2021 to June 2023. Selected images reflected self-expression and emotional resonance with digital fashion. Data was sourced from official brand websites, social media, and online press. Three virtual fashion experts reviewed the dataset, yielding a final selection of 75 images, with Fleiss's kappa coefficients (0.76), confirming substantial agreement (Kraemer 1992). To define affective qualities, 8,031 emotional adjectives were gathered from brand-related online content. Using the Kobayashi Image Scale (1991) and Lee and Kim's (2016) framework, paired adjectives were refined through Multi-Dimensional Scaling (MDS), ensuring conceptual coherence (Cronbach's alpha > 0.9). Further prioritization was conducted through Analytic Hierarchy Process (Vaidya and Kumar 2006), resulting in a final list of 61 emotional adjective pairs, as depicted in Figure 14.1.

A five-point Likert scale evaluation was conducted from October to December 2023, engaging 30 Gen Z fashion design students with expertise in virtual fashion. The results were visualized on a two-dimensional affective quality image scale, mapping relationships between emotional descriptors and design elements along warm-cool and hard-soft spectrums, as shown in Figure 14.2. Further analysis examined

FIGURE 14.1 Emotional vocabulary list of virtual fashion designs.

(Created by authors)

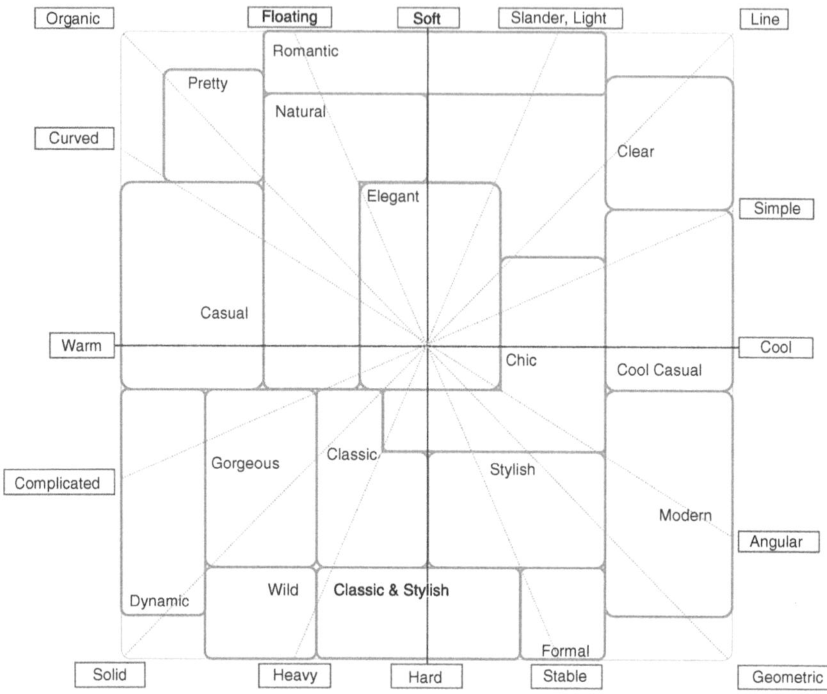

FIGURE 14.2 Affective quality image scale developed.

(Created by authors)

variations across brands and seasonal collections, revealing distinct emotional design strategies. To ensure validity, fashion design educators and digital fashion experts reviewed the findings, confirming their practical relevance and implementation in virtual fashion design.

14.4 RESULTS

Among 61 pairs of emotional adjectives, 17 pairs (27.9%) demonstrated a significant impact, exceeding 70% prominence with a 40% gap over their antonyms. When analyzed by design elements, color (38.9%), visual texture (36.4%), and tactile texture (30.0%) were found to contribute substantially to the affective experience, whereas shape (13.6%) exhibited the least impact. In contrast, 24 adjective pairs (39.3%) had minimal influence, with prominence levels below 60% and gap differences under 20%, further highlighting the greater emotional significance of color and texture in virtual fashion compared to shape. Figure 14.3 illustrates the varying influence of design elements on affective quality in virtual fashion, highlighting their role in shaping users' perceptions and engagement.

Shape was most strongly associated with 'three-dimensional' (76.3%), 'trendy' (73.5%), and 'exaggerated' (71.4%), reinforcing dynamic and chic affective qualities. In contrast, adjectives such as 'standard' (50.2%) vs. 'experimental' (48.8%),

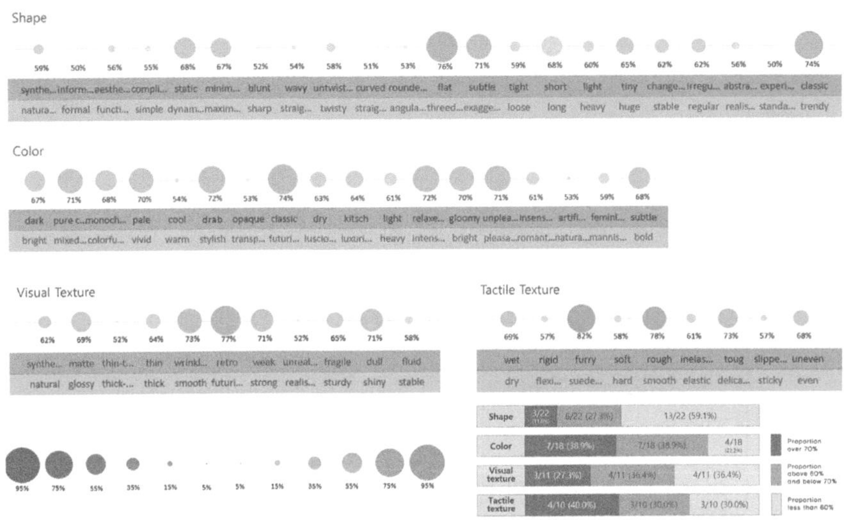

FIGURE 14.3 Distribution of emotional vocabulary by design elements. **(Created by authors)**

'informal' (50.4%) vs. 'formal' (49.6%), and 'curved' (51.4%) vs. 'straight' (48.6%) exhibited minimal differentiation, indicating a limited role in shaping emotional perceptions. Color played a dominant role in shaping emotional responses, particularly through 'futuristic' (74.5%), 'stylish' (72.2%), and 'intense' (71.6%), which strongly influenced modern and dynamic quality perceptions. However, contrasts such as 'artificial' (52.7%) vs. 'natural' (47.3%), 'opaque' (52.8%) vs. 'transparent' (47.2%), and 'warm' (53.6%) vs. 'cool' (46.4%) had a relatively low impact on the overall affective quality of virtual fashion design. Visual texture emerged as a key determinant of modern and stylish emotional expressions, particularly through adjectives such as 'futuristic' (76.9%), 'smooth' (73.3%), and 'strong' (70.7%). Conversely, adjective pairs such as 'thick-threaded' (51.8%) vs. 'thin-threaded' (48.2%), 'unrealistic' (52.4%) vs. 'realistic' (47.6%), and 'stable' (57.6%) vs. 'fluid' (42.4%) demonstrated limited influence on affective quality. Tactile texture, on the other hand, was primarily associated with elegant and romantic qualities, with adjectives such as 'suede-like' (82.3%), 'smooth' (77.5%), and 'delicate' (72.8%) exhibiting the strongest emotional impact. The least impactful pairs were 'slippery' (57.0%) vs. 'sticky' (43.0%), 'flexible' (57.4%) vs. 'rigid' (42.6%), and 'hard' (58.2%) vs. 'soft' (41.8%).

This comprehensive analysis, summarized in Figure 14.4, highlights the relative influence of different design elements on affective quality and their variable contributions across the emotional spectrum of virtual fashion design. Certain elements exhibited greater dominance in specific affective dimensions, reinforcing their role in shaping users' emotional engagement. Notably, the distribution of emotional adjectives demonstrated a strong overlap between color and texture, suggesting that their strategic combination enhances emotional resonance in virtual fashion design collections.

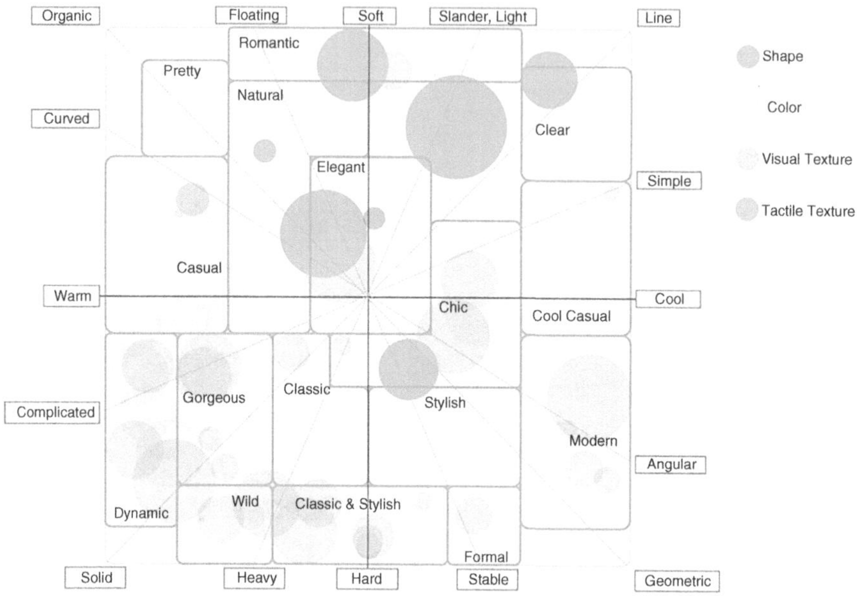

FIGURE 14.4 Affective quality image scale of overall virtual fashion design elements. **(Created by authors)**

Further analysis examined brand- and season-specific variations in affective qualities. Results indicated minimal seasonal fluctuation, with standard deviations of 0.042 for shape, 0.036 for color, 0.051 for visual texture, 0.026 for tactile texture, and an overall average of 0.039. These findings suggest that emotional responses remain consistent over time, indicating that seasonality does not significantly affect affective qualities in virtual fashion. Key emotional adjectives – 'three-dimensional' (shape), 'futuristic' (color and visual texture), and 'suede-like' (tactile texture) – remain consistent across different seasonal collections. Figure 14.5 illustrates this consistency, showing that the distribution of emotional vocabulary exhibited no significant variation in the ranking of the most prominent and least impactful emotional descriptors throughout the study period. However, brand-specific differences were more pronounced, reflecting distinct identity-driven emotional and aesthetic strategies, as shown in Figure 14.6. RTFKT emphasized futuristic and modern qualities reflected in its use of 'futuristic' colors, 'trendy' shapes, and 'smooth' tactile texture. Republiqe adopted a clean and minimalist emotional aesthetics, prioritizing 'bright' color and simplicity in its shape and texture languages. The Fabricant leaned towards emotions under avant-garde styles, characterized by 'mixed' color, 'three-dimensional' shape, and 'futuristic' visual texture expressions. Tribute consistently highlighted futuristic and luxurious textures like 'suede-like' tactile finishes, which enhanced elegant and romantic qualities. DressX, on the other hand, maintained a balance between futuristic and

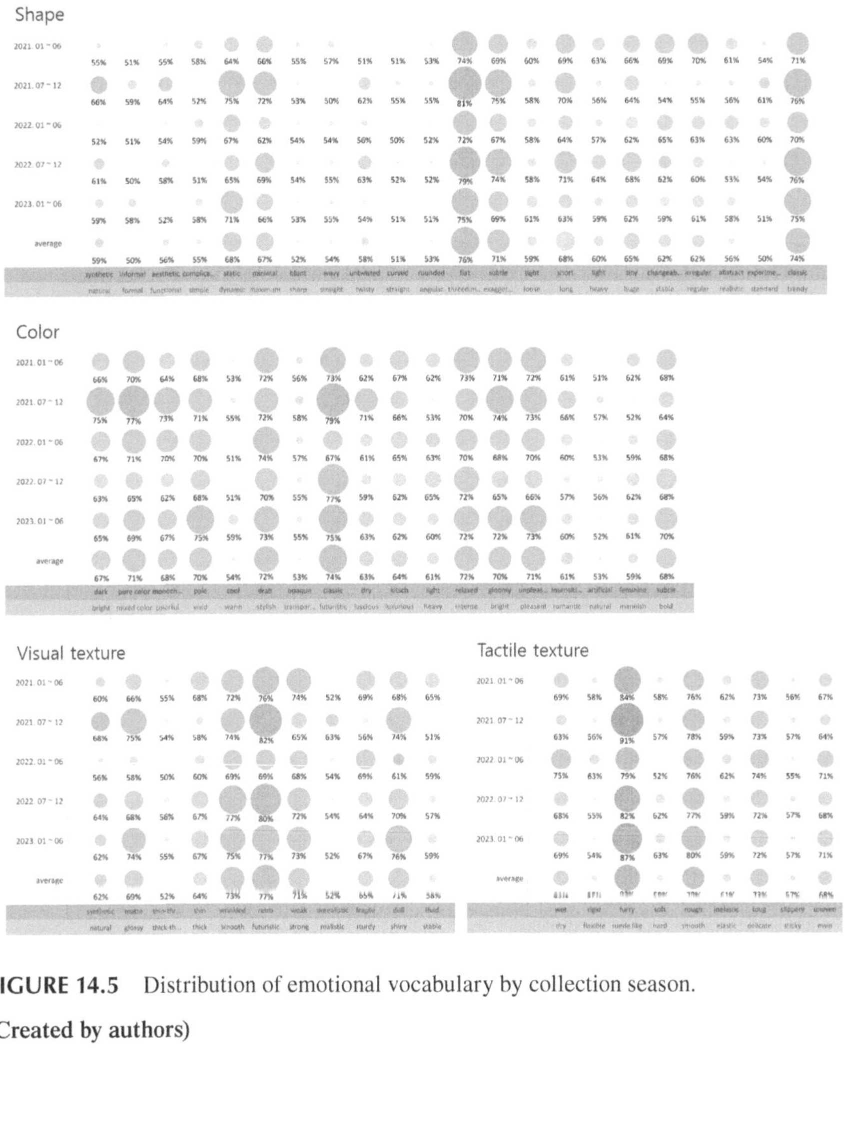

FIGURE 14.5 Distribution of emotional vocabulary by collection season.

(Created by authors)

versatile design expressions via 'stylish' colors, 'stable' shapes, and 'futuristic' visual texture and 'suede-like' tactile texture.

Statistical analysis revealed greater variability across brands (SD: 0.063 for shape, 0.051 for color, 0.072 for visual texture, and 0.048 for tactile texture, and average SD: 0.059) compared to seasonal collections, underscoring the role of brand identity in shaping affective qualities in virtual fashion design. To validate the findings, five digital fashion design experts assessed the relevance and applicability of the study, yielding an average Likert scale rating of 4.21/5 with Cronbach's alpha of 0.98, confirming a high level of reliability.

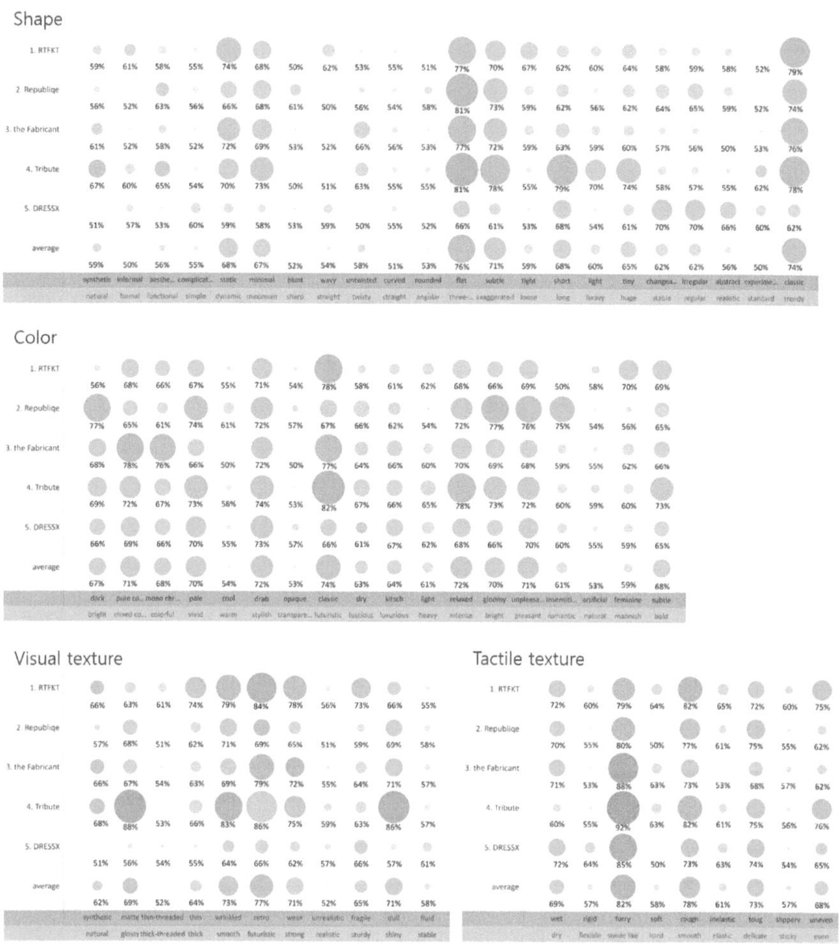

FIGURE 14.6 Distribution of emotional vocabulary by brand.

(Created by authors)

14.5 DISCUSSION: THEORETICAL INSIGHTS AND PRACTICAL IMPLEMENTATION

This study underscored the critical role of affective quality in virtual fashion, elucidating how core design elements like color, shape, and texture influence users' emotional engagement in the Metaverse era. Findings indicated that color and texture exert the most significant impact on user perception and emotional resonance, while shape plays a comparatively minor role. Dominant emotional qualities – including 'dynamic,' 'elegant,' 'chic,' and 'modern' – emerged across virtual fashion collections, reinforcing user engagement and brand differentiation. Notably, the interplay between color and texture was particularly pronounced, with vibrant hues and smooth textures enhancing perceptions of sophistication and elegance, aligning with prior research on emotional design and user experience (Norman 2004; Rathod and Kolhatkar 2014). In contrast, 'pretty,' 'cool,' 'casual,' 'simple,' and 'natural' were less prominently represented. These insights offer valuable implications for virtual fashion designers and brands, suggesting that strategically integrating color and texture can enhance emotional resonance, while identifying underrepresented affected qualities opens avenues for further creative exploration in digital fashion.

Brand-specific analyses confirmed that affective quality serves as a strategic tool for defining brand identity within the virtual fashion domain. While RTFKT prioritized futuristic and modern expressions, Tribute emphasized luxury and elegance, leveraging tactile textures and warm palettes. These variations underscored how brands strategically apply emotional design principles to cultivate distinct consumer experiences, reinforcing their market positioning in the Metaverse era.

Beyond industry relevance, experts highlighted the pedagogical value of this study's findings for digital fashion curricular. Traditional educational frameworks primarily focus on technical competencies such as 3D CAD modeling, often overlooking the emotional dimensions of design. Integrating affective quality frameworks into fashion education would better equip future designers to create emotionally resonant virtual fashion designs, aligning with the evolving expectations of consumers. Furthermore, educational initiatives such as virtual try-on simulations and avatar-based emotional aesthetics could enhance learning outcomes, fostering emotionally compelling digital fashion experiences.

From a practical standpoint, experts highlighted the applicability of these findings in refining marketing strategies and developing tools for emotionally driven designs. As emotional engagement continues to emerge as a critical differentiator in the competitive virtual fashion market, designers and brands must strategically leverage affective qualities to cultivate long-term consumer relationships and enhance digital brand positioning.

While the study offers valuable insights, its generalizability is constrained by the participant demographics, as emotional responses to virtual fashion design may vary based on age, cultural background, and professional expertise. Future research needs to employ broader sampling to capture diverse perspectives on affective quality across various consumer segments. Additionally, experts identified the need for more systematic methodologies to assess affective quality in virtual fashion design, recommending the integration of AI-driven emotional analytics and advanced data visualization tools to further refine emotional design frameworks in digital fashion. Future research should explore how digital innovations, including deep learning and psychological insights, can enhance the depth of emotional design research. Furthermore, expanding investigations into cultural and demographic variations in affective quality will provide globalized design strategies, enabling brands to develop data-driven, emotionally engaging virtual fashion experiences in the Metaverse era.

14.6 CONCLUSION

This study examined the impact of key design elements on affective quality in virtual fashion, advancing both theoretical understanding and practical applications.

Findings highlights that color and visual texture, particularly their synergistic interaction, play a pivotal role in shaping emotional qualities such as 'dynamic,' 'elegant,' and 'modern.' These results underscore the importance of emotional design in virtual fashion and the need for deeper exploration into how design elements collectively influence user perception.

By emphasizing these core design elements, the study advocates for integrating emotional design principles into strategic virtual fashion development. It contributes to the expanding discourse on affective quality while laying the groundwork for future research on the interplay of design elements in creating cohesive emotional experiences of virtual fashion. Establishing a systematic framework for emotional design will be crucial in bridging theoretical insights with industry practice.

Beyond academic contributions, the study provides actionable insights for industry professionals, enabling designers and brands to create emotionally resonant virtual fashion that enhances self-expression and user engagement. Future research should refine these strategies, ensuring that virtual fashion continues to evolve as an emotionally compelling and user-centered medium in the digital economy.

BIBLIOGRAPHIES AND REFERENCE LISTS

Aldridge, Victoria K., Terence M. Dovey, and Angie Wade. 2017. "Assessing Test-Retest Reliability of Psychological Measures." *European Psychologist* 22: 207–218. https://doi.org/10.1027/1016-9040/a000298

An, Hyosun, and Insong Lee. 2015. "A Conceptual Framework for Asian Women's Emotional Needs in Fashion Design." *International Journal of Fashion Design, Technology and Education* 8: 206–213. https://doi.org/10.1080/17543266.2015.1053421

Cascianni, Daria, Olga Chkanikova, and Rudrajeet Pal. 2022. "Exploring the Nature of Digital Transformation in the Fashion Industry: Opportunities for Supply Chains, Business Models, and Sustainability-Oriented Innovations." *Sustainability: Science, Practice and Policy* 18(1): 773–795. https://doi.org/10.1080/15487733.2022.2125640

Hollensen, Svend, Philip Kotler, and Marc Oliver Opresnik. 2023. "Metaverse – the New Marketing Universe." *Journal of Business Strategy* 44(3): 119–125. https://doi.org/10.1108/JBS-01-2022-0014

Jin, Mingya, and Sun Yeong Choi. 2023. "Chinese Gen Z's Emotional Dimensions for Fashion Design during the Pandemic." *International Journal of Fashion Design, Technology and Education* 16(1): 119–129. https://doi.org/10.1080/17543266.2022.2126897

Khelladi, Insaf, Catherine Lejealle, Saeedeh Rezaee Vessal, Sylvaine Castellano, and Domenico Graziano. 2023. "Why Do People Buy Virtual Clothes?" *Journal of Consumer Behaviour* 23: 1389–1405. https://doi.org/10.1002/cb.2270

Kim, Hyang-ja, and Young-sam Kim. 2014. "Virtuality in Digital Fashion Images." *Journal of the Korean Society of Clothing and Textiles* 39(2): 233–246. https://doi.org/10.5850/JKSCT.2015.39.2.233

Kim, Minsun, and Jinyoung Kim. 2022. "A Study on the Characteristics of AI Fashion Based on Emotions – Focus on the User Experience." *Journal of Fashion Business* 26: 1–15. https://doi.org/10.12940/jfb.2022.26.1.1

Kobayashi, Shigenobu. 1991. *Color Image Scale*. Tokyo: Kodansha International.

Kodzoman, Duje, Alenka Pavko Čuden, and Vanja Čok. 2023. "Emotions and Fashion: How Garments Induce Feelings to the Sensory System." *Industria Textila* 74(3): 346–355. https://doi.org/10.35530/IT.074.03.202253

Kraemer, Helena Chmura. 1992. "Measurement of Reliability for Categorical Data in Medical Research." *Statistical Methods in Medical Research* 1(2): 183–199. https://doi.org/10.1177/096228029200100204

Kye, Bokyung, Nara Han, Eunji Kim, Yeonjeong Park, and Soyoung Jo. 2021. "Educational Applications of Metaverse: Possibilities and Limitations." *Journal of Educational Evaluation for Health Professions* 18: 1–13. https://doi.org/10.3352/jeehp.2021.18.32

Lee, Jae Min, and Hea-Yeon Kim. 2016. "Study on Sensibility to Visual Tactility of Modern Fashion Materials." *Bulletin of Korean Society of Basic Design & Art* 17(5): 447–460.

Liu, Kaixuan, Shunmuzi Zhou, Chun Zhu, and Zhao Lü. 2022. "Virtual Simulation of Yue Opera Costumes and Fashion Design Based on Yue Opera Elements." *Fashion and Textiles* 9: 1–21. https://doi.org/10.1186/s40691-022-00300-0

Loureiro, Sandra Maria Correia. 2023. "Artificial Intelligence and Extended Reality in Luxury Fashion Retail: Analysis and Reflection." In *Augmented Reality and Artificial Intelligence: The Fusion of Advanced Technologies*, edited by Vladimir Geroimenko, 3230348. Cham: Springer. https://doi.org/10.1007/978-3-031-27166-3_18

Ma Na, Jieun Kim, and Jee Hyun Lee. 2022. "Exploring Personalized Fashion Design Process Using an Emotional Data Visualization Method." *Fashion and Textiles* 9: 44. https://doi.org/10.1186/s40691-022-00321-9

Norman, Donald A. 2004. *Emotional Design: Why We Love (or Hate) Everyday Things*. New York: Basic Books.

Rathod, Ajay, and Avinash Kolhatkar. 2014. "Analysis of Physical Characteristics of Bamboo Fabrics." *International Journal of Research in Engineering and Technology* 3: 21–25. https://doi.org/10.15623/ijret.2014.0308004

Santos, Vasco, Bruno Sousa, and João Simões. 2024. "Metaverse and NFT Market as AI Disruptive Technology for Digital Luxury Brands." In *Business Development via AI and Digitalization: Studies in Systems, Decision and Control*, edited by A. Hamdan and A. Harraf, vol. 537. Cham: Springer. https://doi.org/10.1007/978-3-031-62106-2_77

Stauss, Renate. 2023. "Looking Like a Woman, Feeling Like a Woman, Sensing the Self: Affective and Emotional Dimensions of Dress Therapy." In *Fashion and Feeling: The Affective Politics of Dress*, edited by R. Rilippello and I. Oarkins, 153–176. Cham: Palgrave Macmillan. https://doi.org/10.1007/978-3-031-19100-8_9

Vaidya, Omkarprasad S., and Sushil Kumar. 2006. "Analytic Hierarchy Process: An Overview of Applications." *European Journal of Operational Research* 169: 1–29. https://doi.org/10.1016/j.ejor.2004.04.028

Venturini, Alice, and Martina Columbano. 2023. "'Fashioning' the Metaverse: A Qualitative Study on Consumers' Value and Perceptions of Digital Fashion in Virtual Worlds." *Journal of Global Fashion Marketing* 15(1): 6–22. https://doi.org/10.1080/20932685.2023.2234918

Won, Jong-shin. 2023. "Influence of Formative Principles of Virtual Influencer on the Ads, Brand, and Product Attitude: Mediating Effect of Likeability." *Journal of Digital Contents Society* 24(8): 1675–1685. https://doi.org/10.9728/dcs.2023.24.8.1675

Xu, Sheng. 2023. "The Consumer Behavior Analysis of Virtual Clothes." *Telematics and Informatics Reports* 10: 100047. https://doi.org/10.1016/j.teler.2023.100047

Yang, Eun Kyoung, Jeongmin Kho, and Ruobing Li. 2025. "Affective Qualities in Virtual Fashion: Exploring Designer Perception of Core Design Elements and Emotional Resonance." *Archives of Design Research* 38(3): 107–129. https://doi.org/10.15187/adr.2025.08.38.3.107

Zaki, Yasmina, Meng-Shan Wu, Allam Hamdan, and Jeremy Thomas. 2024. "Fashion Brands in the Metaverse: Redefining Digital Marketing and Branding Landscape." In *Business Development via AI and Digitalization: Studies in Systems, Decision and Control*, edited by A. Hamdan and A. Harraf, vol. 538. Cham: Springer. https://doi.org/10.1007/978-3-031-62102-4_19

15 Requirements to Enable Future "METAVERSE" Revenue Streams for Emerging Fashion Designers

Zowie Broach, Dawn Ellams, Eleanor Rockett, Andrew Bravin, and Philip Delamore

15.1 INTRODUCTION: BACKGROUND

The fashion industry's engagement within gaming has largely been driven by established brands using the digital interface as a new means to reach customers ensuring brand identities transition into multiverses, commonly known as the metaverse. However, for emerging fashion designers, routes into digital markets through gaming interfaces are challenging, with no clear frameworks for revenue streams.

Research presented within this chapter aimed to explore the convergence of fashion and games, questioning how fashion designs can be monetised through video games to create new approaches to fair revenue streams for emerging fashion designers. The research undertaken formed part of a larger Proof of Market study funded through the Research and Development funding stream within the UKRI project; Future Fashion Factory – Digitally Enabled Design & Manufacture of Designer Products for Circular Economies (grant reference: AH/S002812/1).

Research was undertaken within a cross-disciplinary empirical study encompassing law, gaming and fashion design to develop an integral understanding of the contracting and intellectual property (IP) protection requirements, fashion design process and gaming market. A UK-based games company and three emerging fashion designers each with varying degrees of interaction with the metaverse informed the research. Results discuss the key stages required for the initial development of frameworks that will enable emerging designers to access gaming revenue streams, these being (i) Designer IP, (ii) Platform Model, (iii) Democratising Creative IP.

15.2 METHODOLOGY

To develop an integral understanding of the metaverse market and explore the potential for collaborative working relationships across fashion and video games through

 DOI: 10.1201/9781003594482-19

value creation and future revenue streams, the study was undertaken across three key stages:

Stage 1: Existing research was drawn on to contextualise the landscape of fashion and gaming and to scope the current opportunities for designers entering the industry.

Stage 2: A series of interviews with emerging fashion designers, who had varying degrees of interaction with the metaverse, was undertaken to gain understanding for awareness of contracting and IP protection requirements. In parallel, the game company presented an overview of practices to the RCA research team, and interviews were undertaken with the games company fashion design lead, 3D Character artist lead and a representative from the legal department.

Stage 3: A series of analyses were made to understand current approaches to IP and contractual and licensing agreement across the industry.

Insights from the research were used to inform skills requirements for emerging fashion designers to be able to create 3D fashion assets, as well as the IP requirements that will ensure protection of a designer's creative works within a games context and support the development of new revenue streams within the gaming industry for emerging fashion designers.

15.3 RESULTS AND DISCUSSION

15.3.1 Fashion and Gaming as Collaborators

The literature exploring the relationship between fashion and games has been growing since 2018 but is mostly based on marketing, with a distinct gap in research looking at business models (Fu & Liang, 2022). Across industry and academic literature, there is an emergence of research and discussion around embedding digital technology within design, development and collaboration (Choi, 2022). In Wanick et al.'s The State of Fashion Retailing Post-Pandemic, the authors outline that "a process of digital consumerism is emerging from the constant collaboration between brands, games and virtual worlds. Games, for example, are no longer a space solely driven by entertainment [and] brands are constantly engaging and collaborating with game worlds" (Wanick & Bazaki, 2023).

The term 'metaverse' was originally coined in *Snow Crash*, first published in 1992, as describing a 3D virtual world that was inhabited by avatars (Stephenson, 2022), in recent years it has risen to the surface and gained considerable attention as a term that describes a network of 3D virtual worlds bringing together elements from physical and digital realism (Ravenscraft, 2023). Park and Lim demonstrate the significant research gap in research in fashion and the metaverse, especially in areas outside of a marketing focus (Park & Lim, 2023). There are various ways in which brands are beginning to leverage the metaverse, most often to support their marketing strategy through brand experiences that are immersive for the consumer and the offer of digital products. Within the scope of metaverse innovation in fashion, there is an expansion in what we are beginning to see from games companies with in-game outfits and accessories from brands and unique digital experiences, such as fashion shows and exhibitions hosted by a game's platform (Amed & Berg, 2022).

Hermes was one of the first brands to dabble with digital to engage consumers. During the 2018 annual show jumping competition at Paris' Grand Palais, Hermès released *H-pitch*, its virtual take on traditional horseshoe pitching. Available during the week to download from the App Store and Google Play, the game involved players gaining points for throwing five horseshoes as close as possible to a stake, and environments alternated from a fantastical garden to the staircase of Hermès' storied home at 24 Rue Du Faubourg Saint-Honoré. In 2019, Gucci began adding a selection of games to its existing app in what it referred to as "the Gucci Arcade". With visuals designed in a nod to Seventies and Eighties vintage video games, the Gucci Arcade invited players to interact with its house emblems and best-selling designs. The purpose was to showcase the combining of the past with the present, using contemporary technology to tell stories in an eclectic and pioneering way. In 2019, Louis Vuitton released a capsule collection in collaboration with Riot Games video game *League of Legends*; this was referred to as the first of its kind between a global eSport and a luxury fashion house. The collection followed the creation of special prestige "skins" for one of the stars of the online game, Qiyana, by artistic director of women's creations, Nicolas Ghesquière, and was used to create the character Lightning for Louis Vuitton's 2016 advertising campaign. Ghesquière also designed a "skin" for Senna, another character in the game, revealed in January 2020. Within the *League of Legends*, in-game pop-stars were dressed in Louis Vuitton looks. The collection, that could be pre-ordered, was made up of t-shirts, hoodies, and a biker jacket as well as accessories from keyrings and bag charms to designer bags. Burberry also launched its first online game, *B Bounce* in 2019, bringing the gaming experience to customers globally on Burberry.com. Within the game designed to launch the brand's puffer jackets, players race a deer-shaped character to the moon, using supercharged Thomas Burberry monogram puffer jackets. *B Bounce* was also available to play on the super-size screen inside Burberry's flagship store on Regent Street in London.

Post 2020 and the pandemic, Balenciaga released their fall 2021 collection in the form of online game, *Afterworld: The Age of Tomorrow*. In doing so, Balenciaga became the first fashion brand to produce a digital game; a collaboration across consultants, games developers and relevant technological skills. The result was a post-apocalyptic game, set in the near future of 2031, where players moved through different zones from a Balenciaga store to an enchanted forest, meeting avatars wearing different 'rendered' looks from the collection as they move through their 'hero's journey'. Whilst this was a significant breakthrough moment across fashion and games, the games component was not interactive, and, where traditionally in games there would be secrets or collectibles, this interface was more about positioning the brand for the future potential rather than providing a truly immersive gaming experience.

In a move away from large fashion houses driving collaborations within the gaming sector, in 2022 Joshua Kane developed the video game *A Tailors Journey*. Kane's motivation demonstrates the opportunity for less established and emerging designers to engage with gaming from a creative and fashion designer driven perspective.

15.3.2 Emerging Fashion Designers Perspective

The UK has a well-established group of micro and Small Medium Enterprise (SME) fashion design entrepreneurs linked with the UK's arts education system. These entrepreneurs take risks, create innovative prototypes and are globally recognised (Eckert et al., 2022). It could be said that the education received from leading arts institutions in the UK gives designers more of a change of success through close links to industry leaders (Gornostaeva, 2023). Yet, these micro and small brands are limited by their internal capacity, limited external funding routes and the accessibility of production networks and sales outlets (Marsillac & Roh, 2014; Caniato et al., 2009). Such brands are often reliant on their self-marketing, social media exposure, or industry connections to gain traction in the early stages alongside providing quality craftsmanship, uniqueness and innovation within their product categories (Caniato et al., 2012).

To understand potential revenue streams for emerging designers, a series of interviews were undertaken to capture insights from emerging designers. Three emerging designers were interviewed, transcripts from the interviews were then separately coded by two researchers, and the results were compared for validity through adopting the open coding method (Strauss & Corbin, 1998). The raw data that emerged was grouped into subcategories. Key themes that emerged from these interviews were as follows.

Design tools. When asked about the use of digital tools in their current processes, each of the designers had differing responses. One designer commented that a continuum of open-source software is integral for growth in the digital, explaining

> open source is like the building blocks, . . . an amazing resource that should stay free and accessible and same for the building blocks of chunks of code that just do that action, because they're almost like the Lego. It's a bit like buying pens

Design for games. All designers stated that clothing within games form part of a wider identity and feel that designers would further include gender choice, performativity and cultural nuance. To demonstrate the importance of embedding style and designer intuition into the process, one designer explained

> The Sims years ago they used to have a randomize button. . . . So it would be quite great if you had, say five designers in the same game and you could say pick or randomize, maybe there was a free option to see and you get something really interesting.

Design process. In reference to a more classical fashion design approach, one designer referred to their, using the term "digital garbage", and how, through collaboration, they were able to use excess designs that were not included in their physical collections, for different purposes within the growing online world of digital fashion: ". . . my leftover was also files, so it was kind of like digital garbage & I was not using it, so I looked up to what I could do with all the designs." Relative to the area of designing a game, one designer

> discussed embedding some form of 'restriction' into the digital for fashion designers; an element that could be linked to their physical collections, which are often limited by budget and deadlines.

All the designers emphasised the importance of merging methods of creation, between the physical and digital realms, indicating that although there is much potential, the space that interrelates the digital and the physical, that extension of reality, should not be denied. With more than one designer explaining that maybe economically the digital could be seen to assist the physical more, one designer stated "there is still so much to be done with the physical world I feel like the digital is . . . like an appendix."

Design IP and monetization. Designers felt that there was little or no information or full understanding about the details of how a contract would be built and what it might need to cover. Within the area of pricing and monetisation in digital, all designers agreed that different routes or options are needed depending on the design or project. They felt clear steps and systems should be in place for emerging designers to be able to monetise their designs effectively and would consider an industry-approved contract template to manage future collaborations and agreements. One designer explained,

> specific outputs of projects as a final piece, and that means the whole file . . . not just the aesthetic of it, how it's done, etc., the whole file, I feel that must be protected. In between sort of like a service and an object.

15.3.3 Games Company Perspective

To understand the current approaches to fashion asset development within the metaverse, the study identified current practices and processes used within a UK-based games developer. The games developer presented an overview of practices to researchers, and the research team held interviews with the fashion design lead, 3D character artist lead, and a representative from the legal department. Key themes that emerged from these interviews were as follows.

15.3.3.1 Games Developer Fashion System

The fashion team are set up to respond to demand for and supply of the fast-moving digital consumer market within the 3D social world gaming site. At present, they rotate the fashion assets available and exchange shop content through two 'drops' of fashion assets, where new items are added each week, and provide tweaks and updates constantly. The games developer explained, "there are a variety of reasons that a user would spend physical money within a game from loss aversion, exclusivity, competitiveness, impatience or for investment for future benefit."

15.3.3.2 Fashion Assets

The majority of the in-game fashion content (assets) published is generated in-house by company creatives. As a by-product of this advance, there has been a development in IP rights, including house brands and signature looks. On occasion, the games company will outsource content; typically, designs will be licensed from designers or third parties, which is done based on standard terms for a flat fee.

15.3.3.3 Designing Beyond the Fashion Asset

The games company approach to fashion design relies on not only creating fashion assets but also in designing identities, 'single beings', inclusive of anatomy, face, skin, posture and even the emotional status of the 'being'. The fashion assets also include the designing of make-up, tattoos, hair and furniture, providing a constant flow of ideas and aesthetic choices for the gamer.

15.3.3.4 2D to 3D

The fashion team works in a traditional 2D development process for fashion assets. The lead fashion designer then works closely with the 3D character team lead to create the designed assets in a 3D format ready to be dropped and purchased by players for virtual worlds, virtual aspirations, and, at times, virtual experiments within the game. This collaborative process, developed by the games company's fashion and 3D artist teams, demonstrates the skills required by fashion designers to produce 3D assets. The games company 3D character artist explained that some designer ideas just aren't possible, as they are too complex to create within the digital parameters. The games company fashion designer expanded on this by explaining that in understanding what isn't possible, their design team can start to understand what is possible and think in 3D translation mode as they design in 2D.

15.3.4 Intellectual Property Perspective

Intellectual property (IP) is an umbrella term for several rights that exist under law (in both UK and other countries around the world), which includes, for the purposes of this study, the following individual rights copyright, trademarks, designs. Each of these IP rights are rights which are created by human intellectual effort, and they exist to protect the expression of an individual's efforts being exploited by a third party without permission and to reward creativity with a period of exclusive exploitation rights. Emerging designers might not recognise the importance of protecting the intellectual property they create and how such rights can stop other people misusing their creative output.

Designs are a key intellectual property right in the fashion world. Designs protect the shape, appearance or configuration of an article or product and so are highly relevant for articles of fashion as they could, for example, protect the shape of a dress or coat. There are both registered designs (which you apply for) and unregistered designs (which subsist automatically). It is strongly recommended that key product designs are registered by designers as a registered design as an effective tool for preventing others from copying it. Infringement of a design occurs where, without permission from the proprietor, a product in which the design is incorporated, or to which it is applied, is copied. This could include digital representations of clothing designs used in a commercial context.

Trademarks are signs (words/logos) capable of graphical representation that distinguish or identify goods and services. Consequently, trademarks have enormous value in fashion, and famous examples include the brands Burberry, Gucci and logos, such as the Nike swoosh. Indeed, much of the current fashion industry is brand driven, leading most designers to have multiple trademark registrations for their brands, and, as such, there is great value in licensing the use of such trademarks. Trademarks can either be registered (®) or unregistered (™), but since 1994 only registered marks have the benefit of trademark protection, such as statutory rights to use them and stop others using identical or confusingly similar marks in the UK. Trademarks are registered for chosen goods and services, and registration lasts for 10 years and is renewable. If a mark is unregistered, the owner must look to the notoriously expensive and difficult route to prove common law of 'passing-off' to prevent infringement. Infringement of a trademark occurs when a third party makes unauthorised use of a registered trademark in the course of trade. Again, this can be by way of commercial use of trademarks in digital worlds.

Copyright, as its name suggests, copyright is the right to copy and the right to prevent any other person from copying. It protects artistic works (logos, drawings, and paintings), works of artistic craftsmanship (which, in rare cases, may include items of clothing), written works (including software code), drawings, photos, graphics, and musical and dramatic expressions of ideas. Copyright arises automatically as soon as a creative work is made. People sometimes talk, confusingly, of "copyrighting their work", but, in the UK, there is no copyright registration or similar process. IP owners must therefore rely on the use of copyright notices and maintaining adequate records to prove and therefore enforce their copyright. Copyright lasts for the author's life plus 70 years so there is a long period of exclusivity attached to this right. Copyright from a fashion perspective is most relevant for photos, artworks, or drawings that are featured on clothing. UK law has been reluctant to concede that works of fashion qualify as works of artistic craftsmanship as they are viewed as more functional works. Copyright infringement occurs when a copyright is used/copied/reproduced without the proprietor's permission. Again, misuse may occur in digital worlds in addition to physical worlds.

15.3.5 Games and IP

Intellectual property is fundamental to games as a game is, in essence, a collection of various different types of IP. A significant number of contributions go into a finished game, such as the source and object code (copyright), art (copyright), music (copyright), game name/logo (copyright, trademarks), bespoke middleware or other software (copyright, potentially patents), and certain UX features (copyright, design rights).

All forms of IP have their uses in games, with the most commonly used IP rights for game developers being copyright and trademarks. Broadly speaking, copyright covers practically all core elements of the game, including the digital assets, artworks and underlying source code. The digital assets featured in games, for example characters, avatars, and related skins that may feature or use items of clothing and/or fashion, are likely to comprise a number of different rights, including:

- Source code – underlying code for the asset. This could be part bespoke, part licensed from a code library/third party. This will be protected under copyright laws.
- Game art – the artistic look of a character will be protected under copyright laws.
- Character design – the appearance/shape of a character may also be protected as a design right.

To effectively license and exploit the game, the developer needs to ensure that it has ownership or a right to use the different collection of rights. All publishers and investors will want confirmation that developers own all intellectual property rights in the respective games they want published or funded.

15.3.6 Licensing Models

Licensing of intellectual property rights is an important part of both the fashion and games industries. In fashion, where a designer decides to outsource manufacture, distribution, marketing, or collaboration for a collection, license agreements will be required. Such agreements will set out the different intellectual property rights that are granted for use by the other party and the territory, term, and scope for which those rights are used.

In games the licensing may be of relevant rights to third parties such as publishers of a game, but also very importantly of rights that will apply to the bundle of rights that exists in the game. Often games will be built using game engines, such as unreal that license source code or assets from these third parties to feature in their own games.

15.3.7 Licensing and Contracts

Most developers' approach to digital fashion is to either create virtual fashion via its own in-house team or contracts with third parties to create/provide designs. From an IP perspective, IP from employees is typically owned by the employer (under applicable laws) and licensed from third parties, with licenses having varying scope in terms of usage, territory and duration. From an emerging designer perspective, licensing their IP to third parties, such as games developers will be preferable as it means they retain ownership and may use such IP after the license period has expired. They will ideally limit the duration, territory and scope; for example, perhaps licensing use for a single storyline within a game and in a certain territory. Designers may not mind full assignment of IP if the IP in question isn't of personal importance to the designer or their business. This is unlikely to be the case if it is a key design, trademark or a logo that the designer uses or plans to use extensively.

15.4 CONCLUSIONS

The study highlights the requirements to enable Future metaverse revenue streams for emerging fashion designers, as follows.

Designer understanding for parameters of games physic. In terms of either the fashion designer or the gamer, the role of creator in the future will require some level of understanding for game's physics or development of a design tool to support this design translation.

Design skills development for 3D Creation. For emerging fashion designers to access metaverse revenue streams, supporting them to develop a basic level of skill for 3D creation would remove the burden of work from the character artists within games developers, who are often having to amend designs to work in the 3D. It is important to acknowledge the purpose of supporting fashion designers with this new skill should be to enhance the quality and creativity of assets and not the speed at which they can produce more.

Future Fashion and gaming process crossover. The popularity and demand for digital representations of clothing and so-called "virtual fashion" is growing. Fashion education establishments would benefit from encouraging cross-pollination of the fashion and games industries to foster opportunities. It is important for games developers to gain awareness of the fashion design process and the value in what designers do, and for designers to understand how their designs and ideas may practically be introduced into a game.

Designer IP. To be innovative and attractive to emerging designers, games developers may want to consider alternative approaches for engaging designers, including offering a share of net revenue from sales of works, offering additional promotional and/or credit opportunities for designers, offering longer contractual terms of engagement, and/or adopting a model of licensing of IP from designers rather than taking assignment of IP.

Platform Model. Three key revenue possibilities for designers have been identified when designers occupy the following roles: (1) Supplier – License fee, flat fee, revenue share on item; (2) Curator – Revenue from sales of designs and blueprints within the game; (3) Creator – Revenue within the game from items created and sold. Alternative future methods of monetisation for fashion designs within gaming identified include in-game physical stores, seasonality, paid for early access to collections, additional functionality linked to garments, and sale of rare or extremely limited items. Supporting

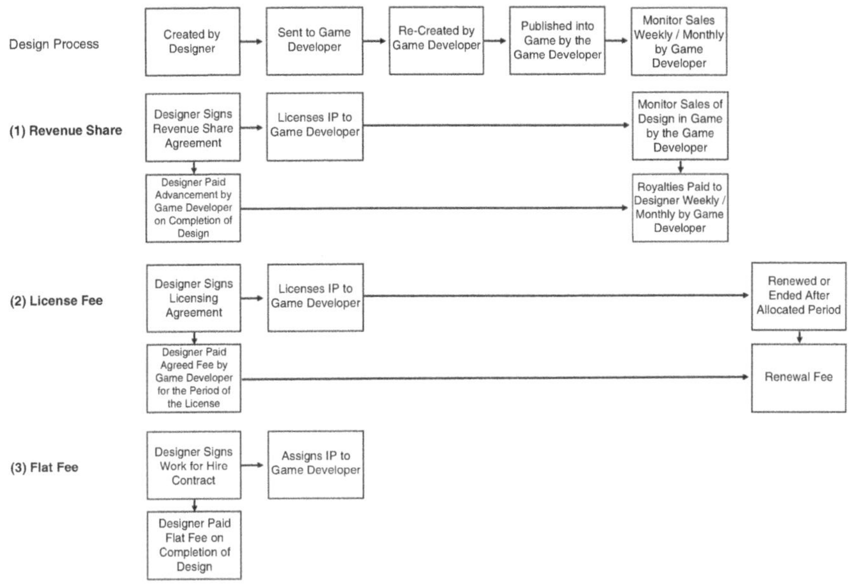

FIGURE 15.1 An outline of the three main options for designers to monetise their IP through a relationship with game developers.

designers to help them facilitate these possibilities is essential. Our research has demonstrated that there are gaps in this "ease" of facilitation for designers. At present, it can cost more than \$200 to put work onto a blockchain if you are an individual.

As seen in Figure 15.1 the options for designers working with a games developer to publish their designs into a game are demonstrated. From this research, it can be concluded that the most favorable agreement for a designer would be a revenue share agreement. Another favorable option for designers is to use a license fee for their designs. The final option of a flat fee through work for hire contract is least favorable as designers assign rights in the work to a games developer.

Democratising creative IP. The idea of an industry standard template for contract and a structured IP process was considered by the designers in interview to be a positive step forward and would be welcomed. The metaverse is likely to be 'metaverses', and any considered agreed structure at this stage will assist in a fair and just protection for users, fans and designers.

Hybrid industries. One of the most central aspects of future hybrid spaces between games and fashion is in the harnessing of emerging fashion designers who have the expertise to explore future identity and have abundance of designs to sell. There is scope to utilise gaming as an operating system for fashion as well as using gaming to diversify revenue streams for fashion designers.

REFERENCES

Amed, Imran, and Achim Berg. 2022. "The State of Fashion: Technology – Download the Special Edition." *The Business of Fashion*, May 3, 2022. https://www.businessoffashion.com/reports/news-analysis/the-state-of-fashion-technology-industry-report-bof-mckinsey.

Caniato, Federico, Maria Caridi, Cecilia M. Castelli, and Ruggero Golini. 2009. "A Contingency Approach for SC Strategy in the Italian Luxury Industry: Do Consolidated Models Fit?" *International Journal of Production Economics* 120 (1): 176–189. https://doi.org/10.1016/j.ijpe.2008.07.027.

Caniato, Federico, Maria Caridi, Luca Crippa, and Antonella Moretto. 2012. "Environmental Sustainability in Fashion Supply Chains: An Exploratory Case Based Research." *International Journal of Production Economics* 135 (2): 659–670. https://doi.org/10.1016/j.ijpe.2011.06.001.

Choi, Kyung-Hee. 2022. "3D Dynamic Fashion Design Development Using Digital Technology and Its Potential in Online Platforms." *Fashion and Textiles* 9 (1). https://doi.org/10.1186/s40691-021-00286-1.

Eckert, Claudia, Philippa Crommentuijn-Marsh, and Sandy Black. 2022. "The Role of Networks in Supporting Micro- and Small-Sized Sustainable Fashion Businesses." *Sustainability: Science, Practice and Policy* 18 (1): 544–559. https://doi.org/10.1080/15487733.2022.2097772.

Fu, Yawen, and Hui'e Liang. 2022. "Sinicized Exploration of Sustainable Digital Fashion: Chinese Game Players' Intention to Purchase Traditional Costume Skins." *Sustainability* 14 (13): 7877. https://doi.org/10.3390/su14137877.

Gornostaeva, Galina, 2023. "The Development of Digital Commerce in the Fashion Industry: The Typology of Emerging Designers in London." In *Technological Forecasting and Social Change*, vol. 186(PA). Elsevier.

Marsillac, Erika, and James Jungbae Roh. 2014. "Connecting Product Design, Process and Supply Chain Decisions to Strengthen Global Supply Chain Capabilities." *International Journal of Production Economics* 147 (January): 317–329. https://doi.org/10.1016/j.ijpe.2013.04.011.

Park, Hyejune, and Rachel Esther Lim. 2023. "Fashion and the Metaverse: Clarifying the Domain and Establishing a Research Agenda." *Journal of Retailing and Consumer Services* 74 (September): 103413.

Ravenscraft, Eric. 2023. "What Is the Metaverse, Exactly?" *Wired*, June 15, 2023. https://www.wired.com/story/what-is-the-metaverse/.

Stephenson, Neal. 2022. "Snow Crash." *Penguin.co.uk*, November 24, 2022. https://www.penguin.co.uk/books/20868/snow-crash-by-stephenson-neal/9780241629833.

Strauss, Anselm L., and Juliet M. Corbin. 1998. *Basics of Qualitative Research: Techniques and Procedures for Developing Grounded Theory*. 2nd ed. Thousand Oaks: SAGE Publications.

Wanick, Vanissa, and Eirini Bazaki. 2023. "The State of Fashion Retailing Post-Pandemic: Trends, Challenges and Innovations." *Reinventing Fashion Retailing*: 1–15. https://doi.org/10.1007/978-3-031-11185-3_1.

16 Digital Wonderland in Web3

Utopian Visions and New Activism within the Gucci Community

Marta Sron and Maxi Heitmayer

16.1 INTRODUCTION

The recent evolution of web3 promises a decentralised digital world (metaverse) heralding greater data ownership and transparency, and empowering individuals as active contributors. The fashion industry, having change at the very core of its DNA (Easey, 2009), recognises the potential of emerging technologies to influence consumer behaviour (Amed & Berg, 2022) and to attract young, digitally-savvy customers (Amed et al., 2019), which is evidenced by heavy investment in the space. With the metaverse projected to evolve into a trillion-dollar market (Bobier et al., 2022; Moy & Gadgil, 2022), brands are prioritising discussions around its potential. Web3 technologies are reshaping the way narratives unfold, emphasising a growing trend of collaboration (e.g. Moncler with Fortnite), which results in the formation of specific digital communities. Discord is one of these "web3 friendly" platforms, and Gucci has begun operations there.

Community, as defined in community psychology, involves a sense of belonging and attachment to a dependable group (Sarason, 1974). McMillan and Chavis (1986) identify four core elements: Membership, Influence, Needs Fulfilment, and Emotional Connection. These elements encompass feeling part of the group, reciprocal impact of group dynamics, meeting diverse needs, and fostering a sense of belonging (McMillan & Chavis, 1986; McMillan, 1996). Community identity is reinforced through symbols and mutual acceptance, facilitating personal growth and connection (Perkins & Long, 2002). Communities, thus, serve as vital support networks, shaping identities and facilitating mutual development. Community psychology further emphasises the importance of envisioning a better future for the group (Watts et al., 2003), which aligns with principles of utopianism and serves as catalyst for societal transformation Levitas (2013) and inspiring progress (Bloch & Adorno, 1988).

While we are still far from the anticipated future scale of this digital revolution, the associated process of business transformation is entering a novel stage

DOI: 10.1201/9781003594482-20

(Meeker & Wu, 2019), and fashion brands develop innovative ways to acquire and retain customers, with early indications suggesting that traditional marketing methods are facing significant challenges in web3 environments (Langer, 2022). This study therefore aims to provide a first, in-depth description of web3 communities centred around a (fashion) brand to guide fashion brands on community development in the web3 space. This study poses the following research questions:

1. How do web3 brand communities work? What are the activities and behaviours exhibited by community members?
2. How do digital communities form within web3 spaces? What characterises and distinguishes them?
3. What are the psychological implications for users of digital communities, and how do these influence the relationships to (fashion) brands?

16.2 METHODS

The study employed a non-participative netnographic approach over three weeks, allowing unintrusive observation of digital communities (Kozinets, 1998, 2002, 2022). Ethical approval was obtained from the University of the Arts London. The Gucci web3 community was chosen for its large membership (53,066 participants at the time of data gathering) and high engagement levels. The dataset consists of approximately 12,000 text and visual entry inputs, each accompanied by its corresponding reactions and comments. Microsoft OneNote served as a Digital Journal to organise data types extracted from the Discord chat, following Baetens (2020) on the importance of images in communication. Thematic analysis was conducted using Braun and Clarke's (2006) framework.

16.3 RESULTS

The thematic analysis of Netnographic data identified three themes with two to five subthemes, respectively (see Figure 16.1). The themes are *collective identity*, *value co-creation*, and *utopia*. All the quotes from community members have been anonymised. In this short paper, only highlights of the findings are presented.

16.3.1 Collective Identity

16.3.1.1 Consumer-Brand Value Fit

16.3.1.1.1 Appreciation of Brand Personality

Members actively discussed the brand's personality attributes, emphasising nouns like "creativity", "luxury", "commitment to quality" and placing special emphasis on innovation. Their sentiments were captured in quotes such as "What a time to be alive, history in the making," and "Gucci is pushin", highlighting their appreciation for Gucci's innovative spirit and excitement. "I am so happy", "Lego Gucci".

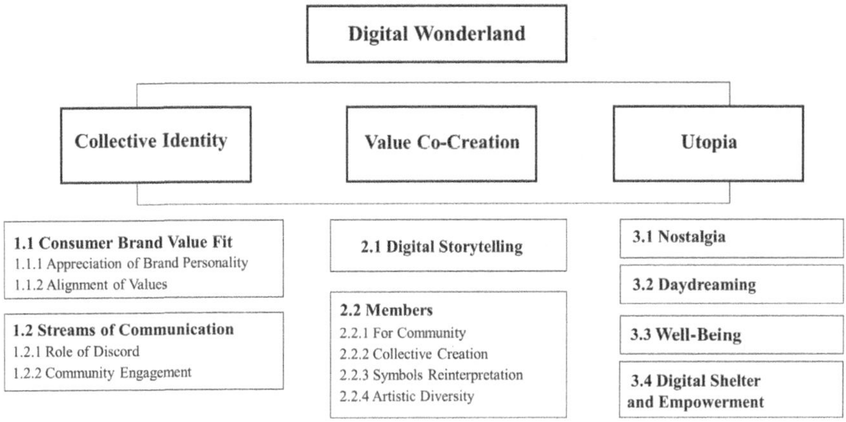

FIGURE 16.1 Overview of themes and subthemes.

16.3.1.1.2 Alignment of Values

Members frequently described Gucci's brand values that align with their own. Comments such as "My favourite aspect of Gucci is their commitment to equality and sustainability" and "Gucci is dope" exemplify this sentiment.

16.3.1.2 Streams of Communication

16.3.1.2.1 Role of Discord

What sets the Gucci community apart from conventional brand engagement strategies is its utilisation of Discord as a communication medium. Discord does not recommend content algorithmically and encourages a more diverse range of expressions like curated memes, videos, special emojis, and stickers than traditional social media. Members chat freely on Discord, where conversations flow naturally due to a transparent format, often with multiple discussions happening at once.

16.3.1.2.2 Community Engagement

The Gucci Discord is a place where members engage in a wide array of conversations about fashion and the Gucci brand, but also casual topics like food recommendations. Furthermore, the community fosters an environment where individuals feel comfortable discussing personal matters like career changes and the challenges of daily life. Additionally, the community regularly highlights that they are part of the group by using collective pronouns ("we", "us"): "What an amazing community we have", "We are Gucci family", "The community matters to us!" Moreover, members refer to themselves as "Gucci Gang" and proudly proclaim their association as part of the "Gucci family".

16.3.2 Value Co-Creation

16.3.2.1 Digital Storytelling

Gucci introduces digital innovations to the community through links to immersive experiences on platforms like *Roblox* (Figure 16.2). These mini Gucci-themed

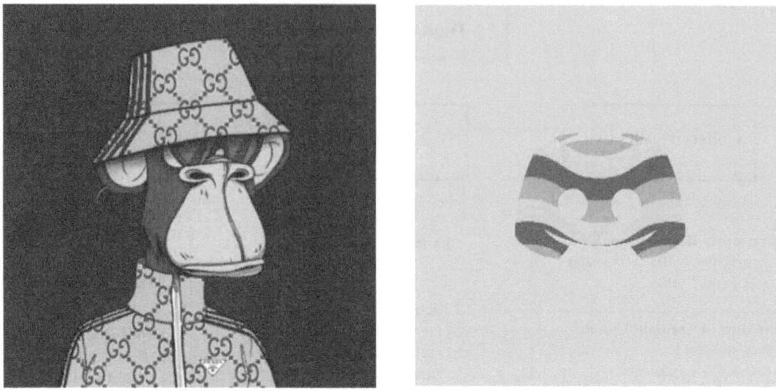

FIGURE 16.2 Profile pictures on Discord showing affiliation to the brand.

FIGURE 16.3 Digital GG World.

worlds allow members to explore the brand's storytelling in a unique digital dimension, often using avatars for navigation (see Figure 16.3).

16.3.2.2 Members

Community members spontaneously create diverse forms of art inspired by Gucci symbols and products, including images, animations, videos, and music, aiming to enrich the collective experience and identity beyond mere brand celebration.

16.3.2.2.1 For Community

One of the examples is the collaborative effort of members who created Gucci stickers, designed to be used as reactions by all members during their chat interactions (see Figure 16.4). This exemplifies their dedication to enhancing the community's

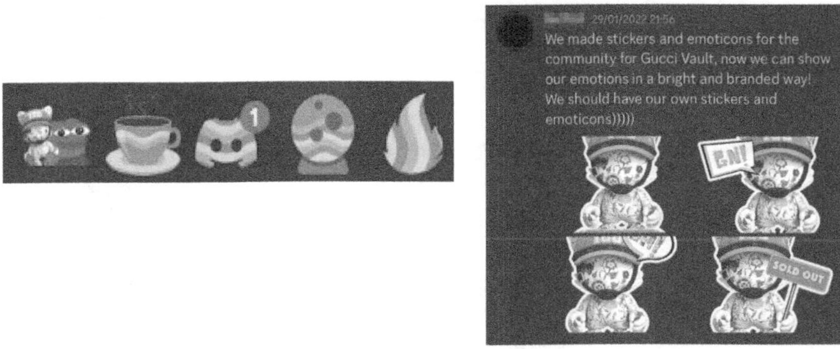

FIGURE 16.4 Example of members' reactions stickers.

FIGURE 16.5 Collective art creation process.

shared experience through the collective enjoyment of creativity. "From community for community".

16.3.2.2.2 Collective Creation

The interpretation of brand storytelling is usually subjective, whereas the Gucci Discord community is creating its own collective symbols to communicate with each other. Members collectively create their own symbols and interpretations of brand storytelling, enriching the narrative and fostering a dynamic collaborative art space, as illustrated by the transformation of real Gucci fashion show models into NFT characters shared in the chat (see Figure 16.5).

16.3.2.2.3 Symbols Reinterpretation

Members create in various artistic styles while the source material being reinterpreted remains the same. An example is the revival of the "Gucci Ghost" concept. Members discussed Trevor Andrew, who transformed Gucci bed sheets into a Halloween costume, leading to a collaboration with Gucci. These conversations sparked fan artworks, showing how narratives can evolve.

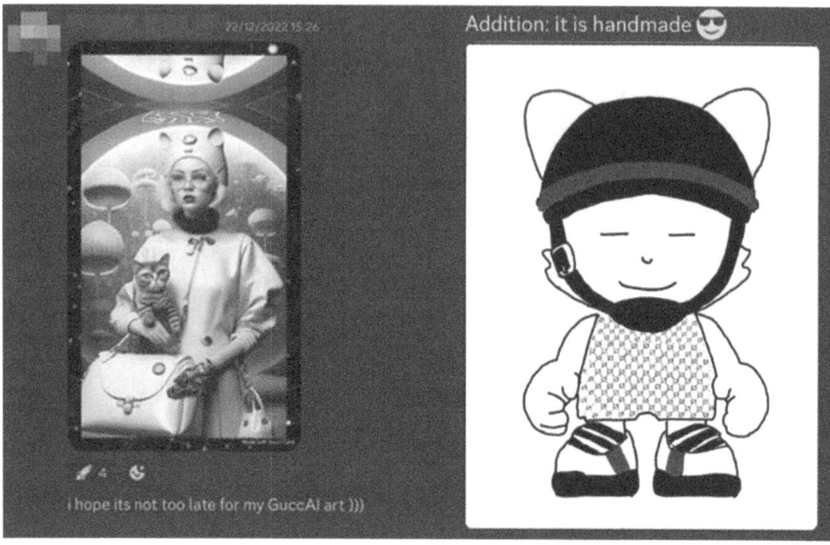

FIGURE 16.6 Examples of various art types created by the community.

16.3.2.2.4 Artistic Diversity

The Gucci Discord Community provides a secure platform to engage in the creation of art centred around Gucci, allowing members to incorporate symbols without legal concerns. However, it is worth noting that art creation within the community attracts all members and is not only the domain of artists. Even individuals who openly acknowledge that they are not artists actively participate in this creative trend, highlighting that the motivation for engaging in these creative activities extends beyond traditional artistic recognition (see Figure 16.6).

16.3.3 Utopia

16.3.3.1 Nostalgia

Within the GG Discord community, members often share stories of their youthful dreams and interests. For instance, they might share childhood aspirations for their dream jobs, reminiscing about the cartoons or shows that ignited their passions. One member shared, "I wanted to become an archaeologist when I was around 8", while another fondly recalls, "That cartoon got me into dinosaurs. I watched it a lot as a kid". These discussions serve as powerful reminders of youthful enthusiasm: "I had a feeling of time travel".

16.3.3.2 Daydreaming

Members of the GG Discord community collectively engage in daydreaming. A recurring theme in daydreaming revolves around fashion, serving as a primary catalyst for imaginative discussions. Members often express their desires in the form of aspirations, such as: "To attend this show physically . . ." or "Must be beautiful in person", and in the form of shared sentiments: "I love floating through space with all of you". Members also frequently envision a future where they can acquire outfits for their avatars in the metaverse, symbolising their forward-looking and technologically driven daydreams: "Imagine buying all the outfits for the avatar in the metaverse".

16.3.3.3 Well-being

There are numerous indicators that the Gucci community is making members feel good about themselves: "Thank you, Gucci fam, you make me happy", "That makes me very happy knowing my work is being enjoyed". Many individuals reported feeling deeply affected by the creative narratives and stories shared within the community. This emotional connection was often expressed through words such as "inspired" and "moved", implying that the community's collective creativity evokes deep emotional responses among its members.

Members of the GG Discord community further exhibited emotional contagion, where one member's expression of joy or amusement triggers a ripple effect of positive emotions among others, which is illustrated, for example, by similar emoji reactions like faces crying with laughter and hearts, which showcases the community's strong emotional connection and shared sense of positivity.

16.3.3.4 Digital Shelter and Empowerment

Members frequently engage in comparing their real-life experiences to their interactions within the GG Discord community (see Figure 16.7). This practice often highlights a perception that the community offers a sense of fulfilment and relief to its members: "Chatting here is a good break for me, work is crazy" and "Feels good", "Whenever you come here you feel Gucci", "I feel that I am finally in a right place". Community fills a role of digital shelter: "Thank you, I can express myself freely here", "I am getting bullied in real life", and "Here, people understand me

FIGURE 16.7 Example of mutual support in GG community.

and my passion for Gucci". This highlights the vital role of this community as a shelter, offering acceptance and understanding that may be lacking in the offline lives.

In this context, members of the GG Discord community consistently offer strong support to one another, particularly when sharing their creative work. They respond with positive comments, emojis, and words of encouragement, creating a nurturing environment: "I love it", "This art blows my mind". The community thus also served as a platform of empowerment for its members.

16.4 DISCUSSION

The current study employed a qualitative netnographic research approach to explore the complexities and nuances of digital communities, particularly focusing on the Gucci Discord community. Given the exploratory nature of the study, the research questions focused on (1) understanding the structure and function of Web3 digital communities, (2) characterising these communities distinctiveness, (3) investigating the psychological implications of being part of such digital communities, and (4) examining the specific activities and behaviours exhibited by community members. The thematic analysis of netnographic data has uncovered deep insights into members of the Gucci Discord as well as the community's dynamics.

16.4.1 Understanding the Gucci Community

The Gucci Discord community differs from traditional online communities, as members are drawn together by both appreciation of the brand personality (Mills et al., 2022) and their personal values alignment with Gucci's (Kristof, 1996; Bhattacharya & Sen, 2003), constructing a coherent social identity (Heider, 2013). However, contrary to previous studies indicating information or entertainment seeking as primary reasons to join online communities (Brown et al., 2007; Hartmann et al., 2015), this did not appear as the primary driving force. Instead, the Gucci community is deeply invested in the brand and its values, leading to strong engagement not just with Gucci but also with fellow members who express and share those values. This engagement goes beyond the "what" of Gucci, including its products, design, and history, to focus on the "who" – the like-minded members and their shared experiences and values. It is a digital shelter where members express themselves without fear of scepticism or bullying. Membership in the Gucci community fosters diverse psychological associations, forging a collective community identity, in line with Social Identity Theory. Members derive a strong sense of belonging from their association (Day, 2011; Tajfel, 1978), which is illustrated by the profile pictures that members choose, for example (Lee & Ottati, 1995; Simon & Pettigrew, 1990; see Figure 16.2). This way of digitally engaging with the brand is characteristic of web3 with its decentralised and user-centric ethos.

16.4.2 Community-Integrated Brand Engagement

Surprisingly, Gucci community members engage in spontaneous creative processes which they share on the Discord chat. They produce various types of art, in diverse formats. They individually or collectively reinterpret Gucci brand symbols, both visual and abstract, ranging from logos to intangible aspects of brand identity like stories, thereby adding new meaning to brand narratives and creating value for the brand and community. This is in line with research on communities establishing territory by symbols (Perkins & Long, 2002). Art creation is not encouraged by a brand, but occurs in an organic way, which is unique in comparison to co-creation when there is a reward system for member input (Vargo & Lusch, 2004; Mathwick et al., 2008; Wasko & Faraj, 2000). There is also no indication that Gucci utilises or commercially exploits this art in any way, nor does it offer compensation or recognition to members. This spontaneous collective creativity represents a novel approach to both brand engagement and interaction within the digital community, between members.

The creation of art serves a dual purpose: Firstly, it celebrates the Gucci brand, enhancing brand engagement; secondly, it stimulates interactions among members. These interactions cultivate mutual support and engagement, strengthening group cohesion (Cai, 2023). The Gucci community's dynamic thus evolves beyond traditional brand engagement methods. We propose the term Community-Integrated Brand Engagement (CIBE) for contexts where members interact with a brand through the social structure of the community in a creative way. Here, the community comes together not merely as passive consumers of content but as proactive members collectively celebrating a brand in a new way. This collective celebration not only fosters a deep connection with the brand but also shapes a community identity centred around it.

16.5 IMPLICATIONS

The interconnected dynamics of this novel way of brand engagement, where members collectively express themselves creatively, forge a collective identity, and underscore the significance of the community, appears to be a disruptive phenomenon and has been organically witnessed within the Gucci Discord community. There is no scientific evidence of such engagement patterns in other online communities, suggesting that this collective creative approach might be a defining trait of web3 communities. It is therefore crucial to determine in future research if this dual-layered creative engagement is common in web3 communities, especially fashion-focused ones, or unique to Gucci's community.

16.5.1 "We are the Gucci Family": Belonging and Engagement

As expected, engagement within the Gucci Discord community meets fundamental psychological needs for belonging and interpersonal connection. Deep bonds in the community align with McMillan and Chavis' (1986) principles of community, offering a safe digital space for mutual support and empowerment. It surpasses mere brand enthusiasm. Therefore, key psychological implications of community participation relate to the community engagement, bonds, and collective artistic involvement corresponding with CIBE. McAlexander et al. (2002) emphasise the deep connections that can arise from shared experiences around a brand. Within the Gucci community, this connection is manifested not merely through dialogues but as an active immersion in shared digital experiences. Furthermore, within the Gucci community, the longing for belonging manifests uniquely: Apart from conversations, members co-live experiences in the digital realm, collectively watch fashion shows, and, most importantly, engage in collective creativity. These practices indicate that while belonging is an intrinsic motivation, it is deeply intertwined with unique social rituals and shared experiences in this community. Drawing upon the insights from Allen et al. (2021), who posit that traditional definitions of belonging are largely confined to interpersonal contacts, the Gucci community offers a distinctive counterpoint. The community's culture, underpinned by the Gucci brand, significantly influences perceptions of belonging. It is not just about interpersonal relationships; it is also about a collective identity shaped by shared cultural and locational attachments creating a harmonious blend of collective and brand identity. The Gucci Discord community thus emerges as a paradigmatic representation of how modern digital communities seek not only social interactions but an active, meaningful role in the brand's storytelling. Being part of the brand's community with an art creation element that shapes the brand's narrative gives members a sense of significance, as they are active contributors to the brand's evolving story. Additionally, while traditional definitions of belonging revolve around interpersonal contacts, the Gucci community offers a unique blend of personal interactions and shared brand affinity in an innovative web3 space (Allen et al., 2021). This distinctive nature can make members feel they are part of something novel and groundbreaking, likely adding to their sense of mattering. It also puts to evidence that individuals derive significant psychological well-being from feeling valued and significant in social contexts in web3 communities, which can serve as buffers against depressive symptoms and anxiety (Dixon et al., 2009; Strayhorn, 2012).

16.5.2 Well-Being and Artistic Engagement

The literature illustrates numerous psychological advantages of engaging with art. Artistic pursuits have been associated with cognitive resilience, emotional balance, and reduced stress, whether they involve invention, distribution, or merely passive admiration (Gillam, 2018; Huotilainen et al., 2018). Therefore, co-creative processes and shared appreciation of art in the Gucci community elevate both individual and collective well-being and enhance self-esteem (Hogg & Mullin, 1999). This sense of well-being results in deeper emotional attachment to the brand (Sirgy & Grzeskowiak, 2007; Ren et al., 2012), which in turn fosters brand loyalty (Safeer & Le, 2023).

16.5.3 New Utopia in the Context of Web3 Communities

The Gucci community is an example of a modern community rooted in web3 and, surprisingly, carries signs of utopianism. The community can be classified as a utopia because members within this community are driven by a desire for a better future, in line with Levitas' (1990a, 1990b) description of utopian aspirations, with the community's visual creativity reflecting elements of modern utopia (Bloch et al., 1986). Yet, it is important to mention that its members remain seemingly oblivious to the presence of the utopian element in the community. Utopia is a by-product of their interactions rather than a goal they pursue; members of the Gucci community rather seek belonging within different structures that resonate with their modern lifestyles (Macionis, 1978; Bauman, 2013). Consequently, digital communities might have become a preferred safe space, offering solace through shared values and aspirations with like-minded individuals. Utopian thinking influences societal and creative engagement (Badaan et al., 2020; Oettingen, 2012; Fernando et al., 2018). When applying this perspective to the fashion industry, especially within communities like Gucci's, there is a remarkable shift in the balance of power. Nowadays, consumers do not just want to consume; they desire impact, and the emergence of utopian digital communities will further democratise their influence. These acts provide glimpses of a utopian vision, embodying members' aspirations for an enhanced future where a collaborative community essence is elevated and shapes the web3 transformation. The community thus acts as a manifestation of these desires, offering momentary escape (Aldwin & Revenson, 1987), crafting an aspirational narrative while aligning with a beloved brand, and potentially acting as a catalyst for progressive thought Hermand, 1975. This resonates with Bloch's predictions that the concept of utopia would evolve into smaller, more focused micro-utopias as the world develops (Bloch, 1989).

16.5.4 Limitations and Recommendations

This study extends established theories of brand engagement and introduces new perspectives of web3-based community utopias utilising Netnography as a way to gain insight into digital fashion communities (Heinonen & Medberg, 2018). However, concerns exist regarding the reliability of data, as findings may not fully represent broader consumer demographics (Mkono, 2013; Xun & Reynolds, 2010). Generalising findings beyond specific online platforms or communities remains a limitation, akin to traditional ethnography (Kozinets, 2002). Additionally, despite efforts to mitigate bias, the researchers' prior engagement with web3 may influence data analysis and perspectives on the subject more broadly. Further research into web3 brand communities is therefore required.

16.6 CONCLUSION

This paper reports the findings of an in-depth netnographic study looking at the digital Gucci brand community on the web3 platform Discord. The findings suggest that this digital community is a new form of utopia that enables members to engage with the brand as co-creators in an artistic, collaborative social structure that is conceptualised as Community-Integrated Brand Engagement (CIBE). This new way of brand engagement focuses an emerging segment of consumers: digital communities. These findings have significant implications for both fashion enterprises, urging a more collaborative approach in this emerging space, and web3 developers, emphasising the importance of catering to these dynamic digital communities. In the words of Luhmann (1976), "utopias are . . . present futures".

REFERENCES

Aldwin, Carolyn M., and Tracey A. Revenson. "Does coping help? A reexamination of the relation between coping and mental health." *Journal of Personality and Social Psychology* 53, no. 2 (1987): 337.

Allen, Karen A., Mark L. Kern, Caitlin S. Rozek, Daniel M. McInerney, and George M. Slavich. "Belonging: A review of conceptual issues, an integrative framework, and directions for future research." *Australian Journal of Psychology* 73, no. 1 (2021): 87–102.

Amed, Imran, Anita Balchandani, Marco Beltrami, Achim Berg, Saskia Hedrich, and Felix Rölkens. 2019. "The Influence of 'Woke' Consumers on Fashion." *McKinsey & Company*, February 12. https://www.mckinsey.com/industries/retail/our-insights/the-influence-of-woke-consumers-on-fashion.

Amed, Imran, and Achim Berg. "The state of fashion 2023: Resilience in the face of uncertainty." *The Business of Fashion*, last modified 2022. https://www.businessoffashion.com/reports/news-analysis/the-state-of-fashion-2023-industry-report-bof-mckinsey/. Accessed April 30, 2024.

Badaan, V., John T. Jost, Joel Fernando, and Yoshihisa Kashima. "Imagining better societies: A social psychological framework for the study of Utopian thinking and collective action." *Social and Personality Psychology Compass* 14, no. 4 (2020): e12525.

Baetens, Jan. "Words and images." *Comics Studies: A Guidebook* (2020): 193–209.

Bauman, Zygmunt. *Liquid Modernity*. John Wiley & Sons, 2013.

Bhattacharya, Chitrabhan B., and Sankar Sen. "Consumer–company identification: A framework for understanding consumers' relationships with companies." *Journal of Marketing* 67, no. 2 (2003): 76–88.

Bloch, Ernst. *The Utopian Function of Art and Literature: Selected Essays*. Cambridge, MA: MIT Press, 1989.

Bloch, Ernst, and Theodor W. Adorno. "Something's missing: A discussion between Ernst Bloch and Theodor W. Adorno on the contradictions of Utopian thinking." In *The Utopian Function of Art and Literature: Selected Essays*, by Ernst Bloch, translated by Jack Zipes and Frank Mecklenburg, 1–17. Cambridge, MA: MIT Press, 1988.

Bloch, Ernst, Neville Plaice, Stephen Plaice, and Paul Knight. *The Principle of Hope*, Vol. 3. Cambridge, MA: MIT Press, 1986.

Bobier, James F., Ted Merey, Sharon Robnett, Michael Grebe, Jack Feng, Bruce Rehberg, Kate Woolsey, and Jessica Hazan. *The Corporate Hitchhiker's Guide to the Metaverse*. Boston, MA: Boston Consulting Group, 2022. Accessed December 1, 2025. https://www.bcg.com/publications/2022/a-corporate-guide-to-enter-the-metaverse-explained

Braun, Virginia, and Victoria Clarke. "Using Thematic Analysis in Psychology." *Qualitative Research in Psychology* 3, no. 2 (2006): 77–101.

Brown, Jonathan, Andrew J. Broderick, and Nicholas Lee. "Word of mouth communication within online communities: Conceptualizing the online social network." *Journal of Interactive Marketing* 21, no. 3 (2007): 2–20.

Cai, Yan. "Strengthening perceptions of virtual team cohesiveness and effectiveness in the new normal: A hyperpersonal communication theory perspective." *Asian Business & Management* 22, no. 4 (2023): 1649–1682. https://doi.org/10.1057/s41291-023-00225-6. Accessed April 30, 2024.

Day, Amanda. *Believing in Belonging: Belief and Social Identity in the Modern World*. Oxford: Oxford University Press, 2011.

Dixon, Andrea L., Corey Scheidegger, and J. Jeffries McWhirter. "The adolescent mattering experience: Gender variations in perceived mattering, anxiety, and depression." *Journal of Counseling & Development* 87, no. 3 (2009): 302–310.

Easey, M., ed. *Fashion Marketing*. 2nd ed. Chichester: John Wiley & Sons, 2009.

Fernando, Joel W., Nicholas J. Burden, Anna Ferguson, and Yoshihisa Kashima. "Functions of Utopia: How Utopian thinking motivates societal engagement." *ResearchGate*, 2018.

https://www.researchgate.com/publication/322596643_Functions_of_Utopia_How_Utopian_Thinking_Motivates_Societal_Engagement. Accessed April 30, 2024.

Gillam, Tim. "Enhancing public mental health and wellbeing through creative arts participation." *Journal of Public Mental Health* 17, no. 4 (2018): 148–156.

Hartmann, Benjamin J., Caroline Wiertz, and Eric J. Arnould. "Exploring consumptive moments of value-creating practice in online community." *Psychology & Marketing* 32, no. 3 (2015): 319–340.

Heider, Fritz. *The Psychology of Interpersonal Relations*. New York: Psychology Press, 2013.

Heinonen, Kristina, and Goran Medberg. "Netnography as a tool for understanding customers: Implications for service research and practice." *Journal of Services Marketing* 32, no. 6 (2018): 657–679.

Hermand, Jean. "The Agrarian ideology: Fascism as Utopia and hypocrisy." *University of Dayton Review* 12, no. 2 (1975): 13.

Hogg, Michael A., and Barbara-A. Mullin. "Joining groups to reduce uncertainty: Subjective uncertainty reduction and group identification." In D. Abrams and M. A. Hogg (Eds.), *Social Identity and Social Cognition* (pp. 249–279). Blackwell Publishing, 1999.

Huotilainen, Minna, Mimmu Rankanen, Camilla Groth, Pirita Seitamaa-Hakkarainen, and Maarit Mäkelä. "Why our brains love arts and crafts implications of creative practices on psychophysical well-being." *FormAkademisk* 11, no. 2 (2018). https://doi.org/10.7577/formakademisk.1908.

Kozinets, Robert V. "On netnography: Initial reflections on consumer research investigations of cyberculture." *ACR North American Advances* 25, no. 1 (1998): 366.

Kozinets, Robert V. "The field behind the screen: Using netnography for marketing research in online communities." *Journal of Marketing Research* 39, no. 1 (2002): 61–72.

Kozinets, Robert V. "Immersive netnography: A novel method for service experience research in virtual reality, augmented reality, and metaverse contexts." *Journal of Service Management* 34, no. 1 (2022): 100–125.

Kristof, Amy L. "Person-organization fit: An integrative review of its conceptualizations, measurement, and implications." *Personnel Psychology* 49, no. 1 (1996): 1–49.

Langer, Daniel. "Luxury brands feel the challenge to master Web3." *Jing Daily*, last modified 2022. https://jingdaily.com/luxury-brands-master-web3-value-creation-bayc/. Accessed April 30, 2024.

Lee, Y. T., and Vincent Ottati. "Perceived in-group homogeneity as a function of group membership salience and stereotype threat." *Personality and Social Psychology Bulletin* 21, no. 6 (1995): 610–619.

Levitas, Ruth. *The Concept of Utopia*. London: Philip Allan, 1990a.

Levitas, Ruth. "Educated Hope: Ernst Bloch on abstract and concrete Utopia." *Utopian Studies* 1, no. 2 (1990b): 13–26.

Levitas, Ruth. *Utopia as Method: The Imaginary Reconstitution of Society*. Springer, 2013.

Luhmann, Niklas. "The future cannot begin: Temporal structures in modern society." *Social Research* (1976): 130–152.

Macionis, John J. "The search for community in modern society: An interpretation." *Qualitative Sociology* 1, no. 2 (1978): 130–143.

Mathwick, Charla, Caroline Wiertz, and Ko de Ruyter. "Social capital production in a virtual P3 community." *Journal of Consumer Research* 34, no. 6 (2008): 832–849. https://doi.org/10.1086/523291.

McAlexander, James H., John W. Schouten, and Harold F. Koenig. "Building brand community." *Journal of Marketing* 66, no. 1 (2002): 38–54.

McMillan, David W. "Sense of community." *Journal of Community Psychology* (1996): 315–326.

McMillan, David W., and David M. Chavis. "Sense of community: A definition and theory." *Journal of Community Psychology* 14, no. 1 (1986): 6–23. https://doi.org/10.1002/1520-6629(198601)14:1.

Meeker, Mary, and Liang Wu. "Internet trends report 2018." *Kleiner Perkins*, 5: 30.

Mills, Michael, Parsa Oghazi, Mats Hultman, and Andreas Theotokis. "The impact of brand communities on public and private brand loyalty: A field study in professional sports." *Journal of Business Research* 144 (2022): 1077–1086.

Mkono, M. "Using net-based ethnography (netnography) to understand the staging and marketing of 'authentic African' dining experiences to tourists at Victoria falls." *Journal of Hospitality & Tourism Research* 37, no. 2 (2013): 184–198.

Moy, Christine, and A. Gadgil. *Opportunities in the Metaverse: How Businesses Can Explore the Metaverse and Navigate the Hype vs. Reality*. New York, NY: JPMorgan Chase, 2022.

Oettingen, Gabriele. "Future thought and behaviour change." *European Review of Social Psychology* 23, no. 1 (2012): 1–63.

Perkins, Douglas D., and D. Adam Long. "Neighborhood sense of community and social capital: A multi-level analysis." In *Psychological Sense of Community: Research, Applications, and Implications*, pp. 291–318. Boston, MA: Springer US, 2002.

Ren, Yi, Fernanda M. Harper, Sara Drenner, Loren Terveen, Sara Kiesler, J. Riedl, and Robert E. Kraut. "Building member attachment in online communities: Applying theories of group identity and interpersonal bonds." *MIS Quarterly* (2012): 841–864.

Safeer, Ahmed A., and Thao T. Le. "Transforming customers into evangelists: Influence of online brand experience on relationship quality and brand evangelism in the banking industry." *Asia Pacific Journal of Marketing and Logistics* 35, no. 12 (2023): 2947–2964. https://doi.org/10.1108/APJML-12-2022-1018

Sarason, Seymour B. *The Psychological Sense of Community: Prospects for a Community Psychology*. San Francisco: Jossey-Bass, 1974.

Simon, Bernd, and Thomas F. Pettigrew. "Social identity and perceived group homogeneity: Evidence for the ingroup homogeneity effect." *European Journal of Social Psychology* 20, no. 4 (1990): 269–286.

Sirgy, M. Joseph, Stephan Grzeskowiak, and Don Rahtz. "Quality of college life (QCL) of students: Developing and validating a measure of well-being." *Social Indicators Research* 80 (2007): 343–360.

Strayhorn, Terrell L. "Exploring the impact of Facebook and MySpace use on first-year students' sense of belonging and persistence decisions." *Journal of College Student Development* 53, no. 6 (2012): 783–796.

Tajfel, Henri E. *Differentiation Between Social Groups: Studies in the Social Psychology of Intergroup Relations*. Academic Press, 1978.

Vargo, Stephen L., and Robert F. Lusch. "Evolving to a new dominant logic for marketing." *Journal of Marketing* 68, no. 1 (2004): 1–17.

Wasko, Marianne M., and Samer Faraj. "'It is what one does': Why people participate and help others in electronic communities of practice." *The Journal of Strategic Information Systems* 9, no. 2–3 (2000): 155–173.

Watts, Ronald J., Natalie C. Williams, and Robert Jagers. "Sociopolitical development." *American Journal of Community Psychology* 31 (2003): 185–194.

Xun, Jin, and James Reynolds. "Applying netnography to market research: The case of the online forum." *Journal of Targeting, Measurement and Analysis for Marketing* 18 (2010): 17–31.

Part E

Miscellaneous

17 Wired for Innovation – Fashion Design and Electronics

Paul Gough, Grace Hsieh, and Feng-Yu Hsu

17.1 INTRODUCTION

Bringing together fashion design and electronics allows designers to create garments that do things they have never done before: to sense the wearer, the environment, to dynamically change and to be digitally connected. However, these creative riches are hard to realise, as there is a large distance between the disciplines, making it challenging to marry the two. In this chapter we share our experience of teaching a seven- to ten-day workshop on electronic fashion, which has been given annually since 2010 (with a hiatus for COVID) to either third-year undergraduates, final-year undergraduates or master fashion design students. Over these 14 years, we have evolved an approach that addresses many of the challenges and enables teams of students to engage with electronics and technical experts to realise their creations. We will discuss our approach, the electronic materials used, together with insights and pitfalls. Finally, we would like to convey that the benefit of such workshops goes well beyond gaining a measure of technical skills; specifically, it gives students the confidence to use new technologies and extend their creative reach.

17.2 BACKGROUND

Electronics offers fashion designers a wonderful toolbox of capabilities that can transform garments from static, digitally isolated objects to dynamic, data generating and paid up members of the Internet of Things (IoT). Sadly these riches turn out to be exceedingly difficult to realise, though a number of niches such as stage wear (Crawford, 2016), red carpet (Mohamed, 2016) and performance monitoring have found success. However we are far from this entering the mainstream.

Given this obvious challenge in bringing fashion and electronics together, why attempt to have a workshop around such a fusion? There are two principal reasons:

- First, a key element of fashion is experimentation; playing with difficult and new materials is part of being a designer. Such challenges move students well out of their comfort zone and to a place where there is little or no

DOI: 10.1201/9781003594482-22

previous work to guide or constrain them. Managed well, this gives students confidence to experiment with new materials.

- Second, working with remote disciplines involves learning to communicate with experts in other fields, sharing ideas, listening and incorporating their advice to jointly realise a working garment. Building a communication bridge to other disciplines, that themselves don't understand garment design, allows students to broaden their creative skills through interaction with others.

These two points give us the motivation for embarking on an apparel workshop despite the challenges. The next step is to examine in more details the nature of the challenges and how to overcome them.

There is no nice mathematical metric that tells us how far one discipline is from another. But we know that maths is close to physics as they share many concepts and language. If we look at fashion design and electronics, we can, on inspection, observe the following:

- Electronics engineers will talk about op-amps, dividers, power rails, firmware; and designers will talk about weave, weft, stitches, toile. Essentially the two disciplines have almost no common vocabulary.
- The materials that both disciplines use are also fundamentally different. Garments and textiles are robust, flexible, strong, washable; while electronics is generally hard, rigid, fragile and, in most cases, not washable.

So for a workshop to be successful we need to be mindful of these issues and construct the workshop accordingly.

17.3 THE POLYMATH DILEMMA

Given the constraints, we can consider two broad approaches to a workshop.

The first is what we have, in the past, referred to as the "polymath dilemma" (Gough, 2016), where the students learn some technical aspects of electronics. There is potentially a large number of electronic skills that a fashion designer could acquire during a workshop, such as programming a microcontroller, sewing with conductive thread, soldering, connecting sensors and recording data and using wireless communication. Essentially, students would acquire a number of a narrow electronics skills, hence the notion of developing a polymath. The dilemma comes from what skills can be learnt in the relatively short time (seven to ten evenings in our case) during a workshop and that would have a lasting benefit.

The second approach is not to focus on developing electronics skills, but to focus on developing communication with an electronics designer, so that together they can realise the student's design. Both approaches have their merits. However, for our workshops we have adopted and developed the second approach, as we are interested in enhancing designers skills to work in multidisciplinary teams.

17.4 THE FIRST WORKSHOP

Before 2010, we considered running an electronic workshop for fashion students, but the lack of garment-friendly electronics hindered our progress. While the field wasn't dormant – with academic research into garments and electronics ongoing – this work largely overlooked pedagogical implications, with the exception of Coleman et al. (2011). Seymour (2008) offers an excellent survey of relevant electronic garment work from that period.

The situation began to change around 2008 when Leah Buechley began exploring ideas in computational textiles, including the development of microcontrollers, switches and LEDs that could be readily sewn into garments (Buechley & Eisenberg, 2008). Maggie Orth was also generating interesting e-textiles around this time (Orth, 2009). A little earlier, Limor Fried (aka Lady Ada) had established the company Adafruit and was designing electronic components and tutorials to make electronics accessible to a broader maker community. It was the availability of these components that enabled the first workshop to proceed.

The first workshop involved third-year fashion design students. There were two classes, each with about 30 students. One class working in the afternoon and the other working in the evenings. Within a class the students were grouped into teams of three or four. The goal of the workshop over eight days was for each team to design a working garment using a combination of LEDs and sensors. On the final day they were required to do a short presentation explaining their design and show their garment working on a model. Three examiners evaluated the design. There were two engineers who worked in both classes to support the teams. The basic structure of the workshop was the following:

- Introduction to electronics apparel with numerous examples from stage wear and sports. Teams were given their design brief (urban evening wear).
- Using conductive thread, the students designed and created a t-shirt containing five LEDs and a coin cell battery. This was an afternoon exercise, and its aim was to give the students confidence to start to play with simple electronics.
- We described to the students, with working examples, additional components that would enable a garment to sense (e.g., accelerometers, microphones).
- The teams then presented their designs to the group and the engineers. In discussion with an engineer, to ensure feasibility of the ideas, the teams finalised their design.
- Three sessions were then devoted to constructing and debugging the garment with embedded electronics.
- The presentation, in the form of a catwalk and short talk explaining the concept and garment construction, was given to the examiners.

On the presentation day all 17 teams produced a garment, and all worked except one. Examples designs from the workshop are shown in Figure 17.1 A workshop video illustrating students' participation and showcasing their design is available (Kromhout, 2010).

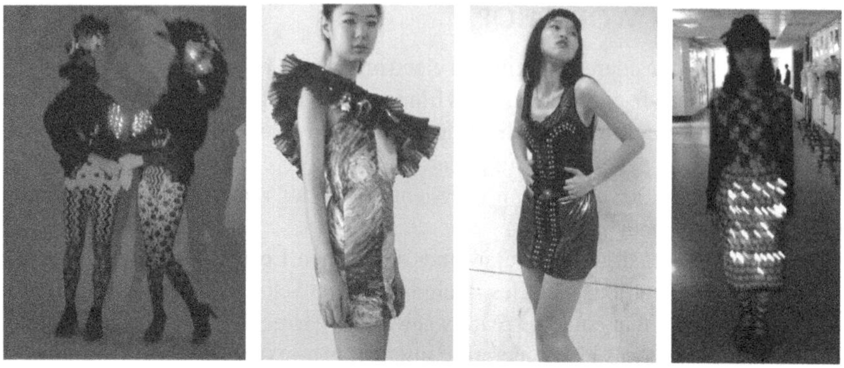

FIGURE 17.1 Examples of student designs from the 2010 Electronic Apparel workshop.

17.4.1 Lessons Learnt from the First Workshop

We learnt a lot of lessons from this initial workshop; the principal ones were the following:

- The students, despite the limitations of the electronics, could produce high-quality designs.
- The approach of having teams of three to four students proved workable.
- We grossly underestimated the amount of effort required to provide the teams with electronic support. Two engineers were not enough; each design involved a substantial amount bespoke design to make it suitable to be embedded in a garment, together with the need for software to control the LED patterns and the inevitable debugging. Failures just before the final presentation also led to some stressful troubleshooting. For future workshop we would need more engineers, though we weren't sure how to achieve that.
- Student feedback and the quality of the designs convinced the department to run the workshop again the following year.

17.5 WORKSHOPS IN 2011, 2012 AND 2013

The workshops for 2011, 2012 and 2013 followed a similar pattern to the first workshop. We overcame the issue of not enough engineers by inviting master-level electronic students from neighbouring NTU (National Taiwan University) to work with the fashion design students. This gives us access to a larger number of engineers. A side benefit of this is that the electronic students also benefit from the workshop, as they gain the experience of working with designers in a creative process.

We continued to practically introduce electronics through the exercise of embedding LEDs and a coin cell holder into a t-shirt. This gave them experience with using conductive thread, particularly, teaching students how to make good contact between components and the conductive yarn. Examples of the t-shirt designs with LEDs are shown in Figure 17.2.

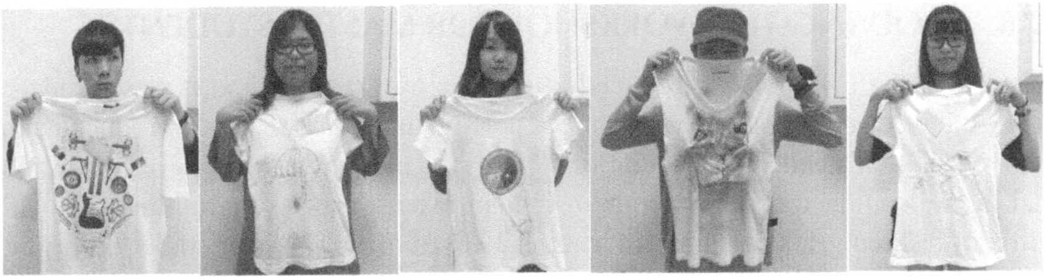

FIGURE 17.2 LED t-shirt from the 2012 workshop.

FIGURE 17.3 Example designs from the 2012 workshop.

Examples of the final student designs for 2012 are shown in Figure 17.3. Although we were generally happy with the workshop output, we were getting more and more frustrated with the use of conductive thread. At first sight using conductive thread seems the most natural material to use with fashion design students; after all, sewing is something they are entirely comfortable with. However, from an electrical perspective the thread has a major problem; it is unshielded, and so unless you are very skilful, it is easy to get electrical shorts or make poor contact to a component. In the workshop this was resulting in hard-to-diagnose problems. Often we observed garments were working when on a table or mannequin but would intermittently fail when worn. Finding these types of failures is difficult and frustrating. We therefore decided after the 2012 workshop to move to using wires, which are stiffer but, due to their shielding, can be safely used in a garment, and failures, such as a broken connection, can be easily identified. Consequently, from 2013 we no longer included the LED t-shirt exercise, as the students didn't need to become familiar with using conductive thread.

17.6 EVOLVING THE WORKSHOP FOR MASTER STUDENTS

After 2013 the university introduced a master's programme, and it was decided to make the electronic fashion workshop a part of that course. This introduced some differences that needed consideration. The most notable was the level of garment making we could expect of individual students. The master's programme is typically a mix of fashion graduates and students that are transitioning from other disciplines, such as architecture, graphic design, etc. The workshop is in the second semester, but some of the students without a fashion design background will only be beginning to develop their garment making skills. Additionally, the number of master students for the workshop is smaller than that from an undergraduate class, typically about 15, which would, in principle, make it possible to have individual projects rather than team projects. We considered two options:

- Individual projects – This would be possible if we allowed students with less garment making skills to take an existing base garment and extend that with their electronics to make a complete design. For example, developing an appliqué containing the electronics would be a possibility to add to a garment.
- Team projects – The alternative is to keep our existing team structure but ensure that each team has a least one person who has good garment making skills. This also has the benefit of encouraging co-working and design, which was valued in the undergraduate workshops. We therefore opted for this second option.

For the master students we introduced preparation for the workshop, as they are expected to be able to develop a theme. Therefore four weeks before the workshop, the students are introduced to a topic which will be the theme for the workshop, for example: Jekyll and Hyde, Extrovert/Introvert, Mood shift. It is based around the idea that a garment with electronics can represent two or more states, and so the preparations goal is to get them to start to play with this idea. The students submit five designs centred on the theme. Each student is then given feedback on the designs and asked to focus in on two designs for elaboration. It is these two designs they take into the workshop and they present to their team members in the early stages of developing a team design. This preparation is done remotely.

17.7 LESSONS LEARNT

We have continued to give the workshop annually, except for a hiatus between 2020 and 2022 due to COVID. The following are some of the key changes we have made to the workshop over the period since 2013 to improve the outcomes.

- Typically we have a ratio of one engineer to two teams. Prior to the workshop the engineers are introduced to the specific microcontroller and programming tool chain that we will use and are expected to have this working on their laptop before the beginning of the workshop. This enables them to focus on working with the fashion design students.

- The initial meeting between the electronic student and fashion design team is quite important. Some engineers confronted with a large number of designers can be a little overwhelmed. So the workshop lead needs to make the introductions, and we have found that starting a discussion on some technical aspect of the team's design gets the conversation underway.
- Don't use conductive thread! I just want to emphasise this for workshops of the type described here, where students are expected to make and produce a full garment during their time. That is not to say conductive threads should be abandoned, but that their use requires the student to develop skills that are difficult to acquire in the timescale of a workshop.
- Students often use multicolour LEDs in a garment. There is a temptation to cover everything in them, making the garment gaudy. We have found constraining the number of LEDs in a garment and getting the students to experiment with covering them and playing with shadow leads to better aesthetics.
- Student expectations have evolved. Smart phones are now a key part of how students access many services and social content, and as such there has been a growing expectation that electronic garments can be similarly accessed and controlled from their phones. Therefore since 2016 we have used a microcontroller with a Bluetooth Low Energy (BLE) capability to enable this. The students generally utilise this capability to control the illumination level, colours and patterns of a garment.
- Ideally the person leading the workshop needs a foot in both disciplines; that is, an engineer that has worked with fashion designers or vice versa.
- Making sure that the students understand that part of the design challenge is usability. Once you are designing a garment that contains electronics, you need to find a comfortable place for the electronics and the battery. In some early designs the battery was left hanging, and someone other than the wearer was required to turn it on! Therefore now we make it explicit that teams will have points deducted if they don't adhere to the following usability requirements:
  - **Placement:** Integrate batteries and electronics comfortably within the garment (e.g., in a pocket).
  - **Safety:** Isolate wires and electronics from the wearer's skin using channels or other protective measures.
  - **Accessibility:** Allow easy access to electronics for repair, debugging and charging. Use snaps and zips instead of permanent stitching.
  - **Control:** Provide a clear on/off mechanism for the wearer.

- Working demos, working demos, working demos. This is a completely new area to the students; they will not have seen electronic garments before, so you need to provide a wide variety of working materials to show and inspire them. In our case we use the following:

  - Historical examples. We are fortunate enough to have examples of electronic garments. Such as the Phillips urban puffer jacket (1998) with electroluminescent display and the Phillips-Levi's jacket (2000). Though these are not working now, the students can take a close look at their construction and gain ideas on how to embed the electronics and displays in a fashion garment. They can also put them on.
  - Students need to understand the variety of available electronic components to realise the functionality and aesthetics they offer, ultimately enabling them to bring their designs to life. For example, we have t-shirts with microphones and motion sensors embedded to illustrate their use. Shaking the t-shirt with the motion sensor causes the LED lights to turn on. Other sensors we use and show are colour, light and temperature sensors. Beyond sensors there are many forms of LED lights that can be used in garments, so these need to be available hands on for the students to play with.
  - Working electronic garments. Examples of garments developed in previous workshops are a valuable source of inspiration, practical help and introduce a little inter-year competition.

Videos of all workshops except 2011 can be accessed (Gough, 2023).

To contrast 'then and now,' in terms of student designs here are examples from the 2018 and 2024 workshops (see Figures 17.4 and 17.5). While the creativity of these garments is similar to past designs, their functionality has significantly advanced. For example, they now include multi-colour LEDs, sensors and are controllable via a mobile phone.

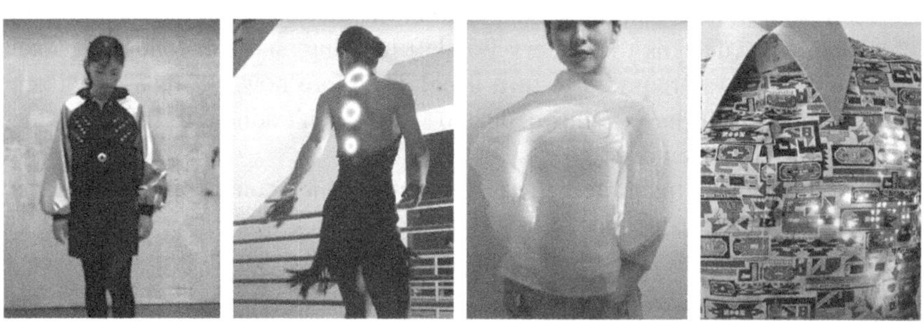

FIGURE 17.4 Examples of designs from the 2018 workshop.

FIGURE 17.5 Examples of designs created during the 2024 workshop.

17.8 STRUCTURE OF THE 2024 WORKSHOP

The current template we use for our most recent workshop:

Pre workshop	
Assignment for student – generate first ideas around design brief	
Induction material for engineers including components that will be used and tool chain	
Workshop	
Introduction to electronic apparel with working examples (workshop lead)	Day 1
Teams meet their engineer	Day 2
Within each team, members share their pre-workshop ideas and generate a team design	Day 2
Once a design is agreed, teams work on generating their working garments	Day 3 – Day 7
Create a short video of the garment capturing its dynamic behaviour and prepare presentation	Day 8
Assessment performed by three examiners – each team presents their of concept, how the garment was constructed, the video and demonstration of the garment on a model	Day 9

17.9 ELECTRONIC COMPONENTS AVAILABLE FOR 2024 WORKSHOP

Figure 17.6 shows the electronic components available to students for their designs:

- Bluetooth low energy microcontroller – All designs contain this component. It is used to read data from connected sensors, control each pixel of any LED strip and communicate with a mobile phone or tablet using Bluetooth.
- LED strips. These are flexible strips of varying density of LEDs from 30 per metre, through 60 to 144 LED's per metre. These happily operate using voltages in the range 4–5V, so they are safe for wearables.

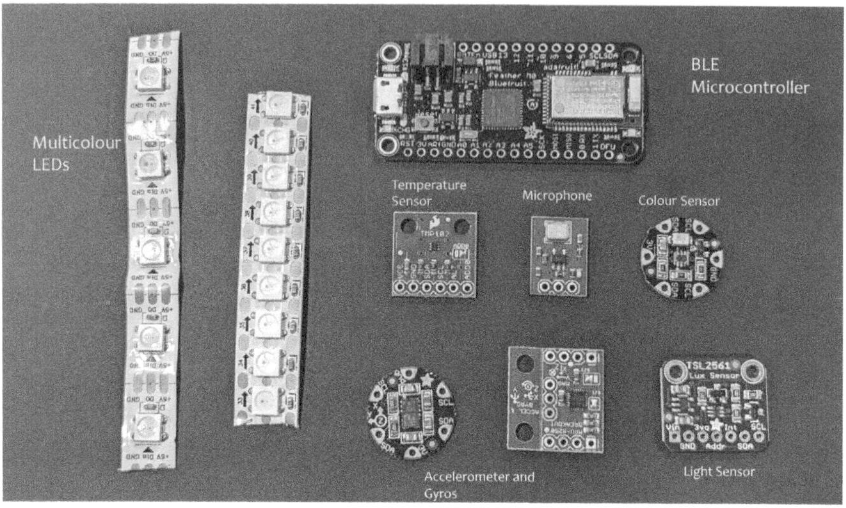

FIGURE 17.6 Examples of electronic components available to students in the 2024 workshop.

- Sensors:
  - Light sensor
  - Colour sensor
  - Accelerometer and gyro – for measuring movement and orientation
  - Temperature
  - Microphone to make sound reactive garments

17.10 OTHER WORK IN THE FIELD

In the intervening years, related work has appeared. We note a similar exploration from Genç et al. (2018) where they investigated the pedagogical implications of integrating computational materials into fashion design by conducting a workshop with fashion design and engineering students, leading to insights into how to teach and inspire wearable design. The university of Lapland has, since 2015 (Harjuniemi & Genç, 2020), run an e-textile course for master interior and textile design students. This adopts a strong interdisciplinary teaching team. The development of designer specific tools to support fashion designers is described by Seyed and Tang (2019). The tool attempts to bridge the gap between fashion and technology by providing a DJ mixer-like interface to prototype interactive garments with light and sensor components.

17.11 INCREASED RELEVANCE TODAY

In the 14 intervening years, electronics for embedded garments has improved in size, variety and flexibility. Initially, we were happy to blink a single-coloured LED; today, we have multicolour LEDs, tiny sensors and garments that can be controlled remotely and connected to the internet. While it is foolhardy to predict, as many have done before, the imminent breakthrough of electronic clothes into the fashion mainstream, it is notably closer as technology is no longer the bottleneck it once was (Ismar, 2020; Muhammad Sayem, 2020). Although there are still some component blind spots, such as reliable electrical connectors for garments, the challenge is now as much about design as it is technology. For instance, multicolour LEDs can easily result in garments that look inaesthetic and appear as novelties. It takes time to tame LEDs dynamic capabilities and for designers to find ways to enhance a garment using them. Thus, the workshop is perhaps now more relevant as we seek to take on these design challenges.

17.12 THE FUTURE WITH AI

AI will, it is anticipated, impact many aspects of fashion design and its supply chains. One aspect that we can explore in the workshop is to use a Large Language Model (LLM) as an expert in some aspect of electronics that students can, through a chat interface, have a dialogue with. In May 2024 we included an LLM (Chat GPT4) as an experiment to support students in developing dynamic LED patterns for a garment, including generating the code to run on a microcontroller. This aspect of the electronic garment design is time-consuming, involving considerable trial and error. Consequently with the workshop time constraints, students neglect this aspect, which can result in poor use of light. Although highlighted in the workshop introduction, the fashion design students preferred, in all cases, to convey the LED patterns they wanted to the engineer, and it was the engineer that used the LLM to help generate the code. Given the workshop's structure of fostering interaction with electronic experts, it is a natural extension to incorporate AI into the framework in this way, but from this initial experience we need to think more carefully about how to introduce it.

REFERENCES

Buechley, Leah, and Michael Eisenberg (2008). "The LilyPad Arduino: Toward wearable engineering for everyone." *IEEE Pervasive Computing* 7, no. 2: 12–15.

Coleman, Melissa, Marloes M. R. Peeters, Véronique Lamontagne, Lone Worbin, and Marina Toeters (2011). "Disseminating Knowledge of Electronic Textiles at Art Schools and Universities." In *Proceedings of the 17th International Symposium on Electronic Art*. Istanbul, Turkey: ISEA, 1–7.

Crawford, C. (2016). "You light up my life: Hacking the senses with wearables." *AATCC*. https://doi.org/10.7939/r3-pd1n-w362

Genç, Çağlar, Oğuz Turan Buruk, Sejda İnal Yılmaz, Kemal Can, and Oğuzhan Özcan (2018). "Exploring computational materials for fashion: Recommendations for designing fashionable wearables." *International Journal of Design* 12, no. 3: 1–19.

Gough, P. (2016). "Fashion design students and electronics – finding a way to work together." *Research Alive*. https://youtu.be/M89IsOMlqOI?si=7oHAfg3izzz0pDFB

Gough, P. (2023). "Electronic apparel workshops." *lab-alpha7.com*. http://lab-alpha7.com/?page_id=35. Accessed January 15, 2025

Harjuniemi, Emmi, and Çağlar Genç (2020). "Arctic and traditional textile techniques as inspiration for electronic textiles." *Tradition and Innovation in Art and Design Education* 178.

Ismar, E., et al. (2020). "Futuristic clothes: Electronic textiles and wearable technologies." *Global Challenges* 4: 1900092. https://doi.org/10.1002/gch2.201900092

Kromhout, F. (2010). "Wearable technology fashion design workshop 2010." *YouTube video*. Posted 2010. https://youtu.be/BJmTdDrqJ5M?si=9hNr6zN_zLCYP5hO

Mohamed, S. S., et al. (2016, April). "Using illuminative technology in creating fashion designs for women evening wear collections." *International Design Journal* 6, no. 2: 109–131.

Muhammad Sayem, A. S., et al. (2020). "Review on smart electro-clothing systems (SeCSs)." *Sensors* 20, no. 3: 587. https://doi.org/10.3390/s20030587

Orth, M. (2009). "100 electronic art years." *maggieorth.com*. http://www.maggieorth.com/art_100EAYears.html. Accessed January 15, 2025

Seyed, Teddy, and Anthony Tang (2019). "Mannequette: Understanding and enabling collaboration and creativity on avant-garde fashion-tech runways." In *Proceedings of the 2019 Designing Interactive Systems Conference*, June 23–28, San Diego, 317–29.

Seymour, Sabine (2008). *Fashionable Technology: The Intersection of Design, Fashion, Science, and Technology*. Vienna: Springer. ISBN 9783211744987

18 "Still Finding its Feet"
Assessing the State of the Digital Fashion Field

Alice Janssens

18.1 INTRODUCTION

While digital fashion had developed during the 2010s to encompass video game-centric "skins" markets and a few luxury brand advertising campaigns, it truly entered public discussions in the early 2020s, fostered by Covid-19 pandemic lockdowns and increasing technological engagement from brands and consumers (Boughlala and Smelik 2024). Year 2022 predictions suggested that Metaverse platforms and surrounding technologies could be valued at between $8 and $13 trillion by 2030, while pickup of digital fashion by gaming platforms such as *Roblox* and major fashion and lifestyle brands including Ralph Lauren and Nike saw significant success (Citi GPS 2022; Monahan 2024; Bain & Company 2023; Keh 2023). But in 2023, the hype and funding for digital fashion and Web3 products waned (Loftus 2024a). Digital innovations such as these have often been regarded as "one offs" or marketing gimmicks by an industry fixated on novelty, despite research emphasizing the benefits that digital fashion can bring to supply chains, sustainability initiatives and consumer engagement (e.g., Van Volkinburg 2020; Periyasami and Periyasami 2023; Colombi and D'Itria 2023; Saxena 2022; Future+ and Luxurynsight 2023; Future+ and Luxurynsight 2024; Ihenacho 2024; Zhang et al. 2023). This belief in digital fashion's gimmick nature was evidenced by the reported 70% year-on-year slump in Metaverse Fashion Week attendance during the event's second iteration in 2023 and the nosedive of the NFT market and related sectors (Hanson 2023; McGimpsey and Leggatt 2024). By 2024, the Business of Fashion and McKinsey & Company's *The State of Fashion* report warned of a turbulent climate for fashion which had "seen a fair share of technological enthusiasm that has swiftly sputtered" (2023, 63).

Despite increasing recognition of effective digital fashion use cases, especially within the context of sustainability, and the growing footprint of events and organizations such as Digital Fashion Week and the Digital Fashion Designers Council, the sector's position remains tenuous (Loftus 2024b; Zwieglinska 2024; Spicher et al. 2024; Digital Fashion Week 2024; Digital Fashion Designer's Council 2024). Threats assuage the fashion and retail industries which struggle with slower growth and increasing uncertainty driven by geopolitical conflicts and shifting consumer behaviour (Business of Fashion and McKinsey 2024). With wary management guiding much of fashion and retail, questions arise about the relevance and lasting

DOI: 10.1201/9781003594482-23

potential of digital fashion. This situation is compounded by the speed at which industry practice and technologies develop. While a growing body of pertinent research addresses the digital fashion sector and its potential to drive positive change within the fashion industry, attaining a broader understanding of the space remains challenging, exacerbated by the competitive funding and publication schedule reliant nature of academia, which moves at a far slower pace. Accounting for these differences in speed and relevance, this study explores the state of play within the digital fashion field, highlighting challenges and opportunities facing practitioners, users, researchers, educators and students.

18.2 DIGITALIZATION, FASHION 4.0 AND BEYOND

While the relationship between fashion and digitalization has received much consideration within scholarly literature, with subjects ranging from mediatization to sustainability and intellectual property, a shift has taken place in the understanding of digital fashion (Rocamora 2016; Black 2018; Janssens and Lavanga 2018). Early analysis focused on design-related perspectives emphasizing the development of technology, processes and images (e.g. Harris 2005; Volino et al. 2005). With technological advancements and the growth of a digitally savvy online public, digital fashion and its study has shifted focus from design to a broader range of topics across fashion making, marketing and management (e.g. Arribas and Alfaro 2018).

Much analysis exploring digital fashion stems from the discussion of "Fashion 4.0", emphasizing the principles of "interoperability", "virtualization", "decentralization", "modularity", "service orientation" and "real-time capacity" (Särmäkari 2021a; Bertola and Teunissen 2018). Fashion 4.0 focuses on the development of new technological, organizational and industrial systems which result in "smart product, production and networks, automation, optimization, flexibility, as well as sustainability-oriented, datafied and customer-driven processes" (Särmäkari 2021a, 88; Bertola and Teunissen 2018). Research addresses new business models, smart garments, AI, the Internet of Things and the use of 3D within design, production, distribution and marketing (Bertola and Teunissen 2018; Mattila 2016; Luce 2018; Farren and Hutchison 2004; Black 2018). The body of literature further explores the growth and increasing relevance of these areas during the Covid-19 pandemic and after. This highlights the development of digital fashion as a tool, space, culture and community which enabled connection and continued creation during periods of lockdown (Särmäkari 2021b; Drinkwater 2024). Case studies investigate the approaches of well-recognised 3D fashion design firms, providing valuable insights into the role of community-based legitimation, shifts in skill development and questions about barriers to large-scale consumption of digital fashion goods (Särmäkari 2021a, 2021b).

Further works examine the role of digital developments in the fashion supply chain highlighting barriers to adoption and digital hesitancy (Heim 2022; Sin Ying and Mok 2023). Much focus has also been given to the use of digital fashion as a marketing tool and the interrelationship between fashion, gaming and identity development (Ayada and Elmelegy 2014; Bendoni 2017; Skjulstad 2018; von Wachenfeldt 2019; Noris et al. 2020; Makryniotis 2018; Park and Chun 2023). The potential of digital fashion to impact sustainability and equity is also explored with reference made to strategies such as virtual sampling, traceability and authentication, production on-demand and virtual-only products. These analyses effectively employ a critical lens which aims to challenge blanket claims that digital fashion and Metaverse systems present "a clean slate" for the fashion industry (Wilczek 2024; Bernat 2024, 214).

Due to growing scientific interest and rapid sectoral development, a range of digital fashion definitions prevail, leading to ambiguity. Baek et al.'s 2022 analysis results in the definition "the virtual creation, production, and representation of one's identity via computer-generated design" (1). Särmäkari's recognition of digital fashion as an "emerging subfield within the field of fashion design, differentiating itself from the professional conventions and building new strategies of jurisdiction and legitimation" emphasises the role of skill differentiation and community (2021a, 85). This is complemented by the clarification of Sayem et al. who provide a useful division of digital fashion into four categories: "Digital fashion design and e-prototyping", "Digital fashion business and promotion", "Digital human and Metaverse" and "Phygital apparel and smart wearable technologies" (2023, 4). This segmenting clearly indicates the breadth of the field and the capacity of digital fashion technologies to bring innovations to the whole length of the fashion value chain. Yet, such divisions can limit a holistic understanding of the state of the sector and its needs. Chan et al. (2023) similarly recognise diversity within digital fashion products, differentiating between digital fashion tools and digital fashion end-products. Despite these clarifications, Spicher et al. indicate that

> defining digital fashion presents some difficulties, not only because as a subset of fashion, it has been around for a mere fraction of the time that fashion itself has, but also because the concept must remain flexible enough to allow new-use cases that we cannot yet imagine.
>
> (2024, 3)

18.3 METHODOLOGY

Addressing the ambiguity around digital fashion's definition and cognizant of the sector's rapid development, this study proffers a view into the structure of the digital fashion field, the opportunities it provides, challenges it faces and its potential for longevity. It is informed by a qualitative methodology, focused on themed collective discussions which account for "the ways in which individuals collectively make sense of a phenomenon and construct meanings around it" (Bryman 2008, 476). This work is based on data collected at the "Digital Fashion in Practice: Applications and Longevity" knowledge exchange workshop held at Winchester School of Art, University of Southampton (WSA) in collaboration with University of the Arts London (UAL) on April 16, 2024. This one-day closed workshop brought together a purposively sampled cohort of six industry practitioners, six academics from the University of Southampton and UAL and 10 students selected by application from the UAL MA Strategic Fashion Marketing, WSA BA Fashion Design, WSA BA Fashion Marketing and Management and WSA MA Fashion Design degree programmes. This physical cohort was complimented by the digital attendance of three industry keynote speakers and six industry provocateurs.

To provide insights from across digital fashion, the workshop was organized into four sessions. The first, "Producing Digital Fashion", explored the creators and landscape of the sector, examining production, events, collaboration, barriers to entry and trends, while the second, "Using Digital Fashion", looked at utility alongside production and strategic applications of digital fashion tools. The third session, "Engaging Audiences", examined requirements for engagement, purpose and platforms, while "Educating for the Future" explored industry needs and how digital fashion is embedded within formal and open-access education. These sessions were complimented by six short digital provocations from industry figures who posed questions to attendees, and two keynote presentations with the first keynote including two speakers (the workshop structure is exhibited in Figure 18.1). Each of the sessions and keynotes were recorded, transcribed and anonymized. The collected data was analyzed with NVivo employing reflexive thematic analysis (Byrne 2021). The analysis of workshop transcripts resulted in the development of key themes, three of which are explored later. These are "Language, Definition and Purpose", "The Digital Fashion Ecosystem" and "Barriers to Growth", presented within their code context in Figure 18.2.

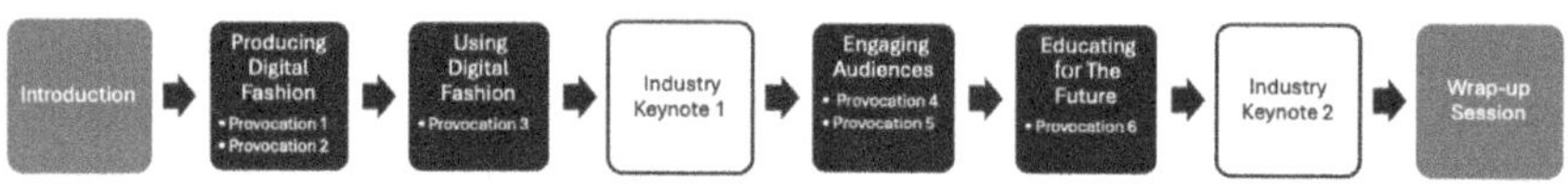

FIGURE 18.1 "Digital Fashion in Practice: Applications and Longevity" workshop structure.

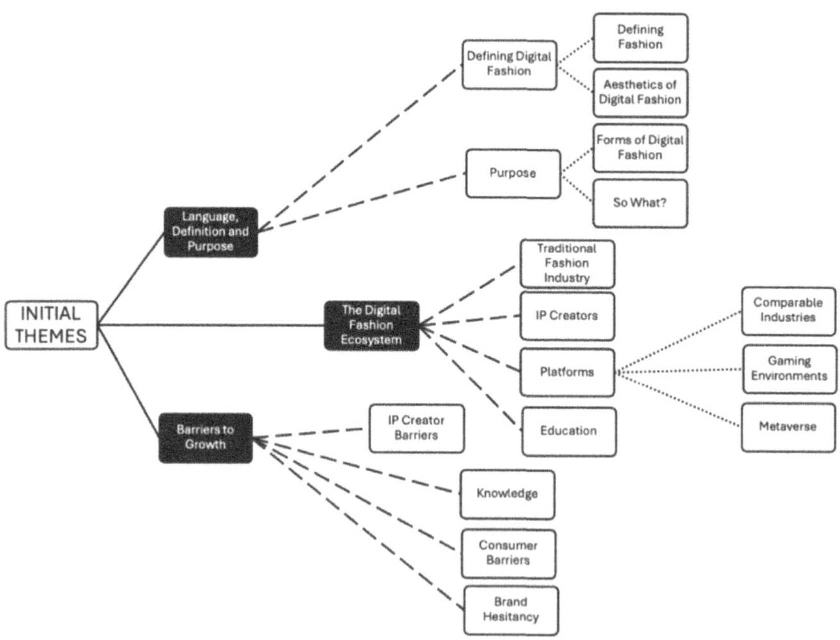

FIGURE 18.2 Initial themes map with three explored codes and sub-codes.

18.4 LANGUAGE, DEFINITION AND PURPOSE

During the planning of the workshop, the organizers chose not to define the boundaries of digital fashion. This decision was made to encourage the participation of figures with diverse practices and to account for developments in the field. This approach proved relevant as much of the first session focused on definition, language and purpose. Participants concluded that these were vital to the sector's analysis and, as suggested in the work of Spicher et al. (2024), ambiguous. Despite much focus in the workshop on the downstream end of the value chain and digital fashion as an end-product, a key point of emphasis was the breadth and diversity of the field. Discussion highlighted the base definition of "fashion" with one participant clarifying

> fashion isn't clothes, garments, utility, necessarily. You know, it's brand, it's storytelling, it's whatever you ascribe to it personally . . . so digital fashion, I think, can be as broad as that. The only difference is that you use digital assets to aid you on that journey.
>
> (Session 1, Participant 1)

A diverse range of digital fashion forms were identified, spanning 3D and CGI design, supply chain developments, digital assets, avatars and metahumans, phygital garments and storytelling and marketing taking place within games, Metaverse platforms and via hologram. However, one participant emphasized that the popularity of many of these aspects within the broader fashion industry is yet to develop, with businesses currently taking "tentative steps" to integrate digital fashion technologies despite potential benefits across the value chain (Session 1, Participant 1).

The diversity of digital fashion forms discussed gave rise to a useful framework where one participant placed use cases on a continuum defined by their risk and limitations to consumers (see Figure 18.3). This participant ascribed higher levels of risk and greater limitations to physical fashion items, followed by phygital items. She clarified that digital fashion items have the lowest level of risk and limitation, suggesting that AR and gaming could provide opportunities to de-risk exploration for consumers who want to "reconnect with themselves without having to engage with the personal, social, environmental and monetary costs that these explorations might have when undertaken physically" (Session 2, Participant 5). There is potential to consider these risks levels for consumers as somewhat parallel to those of producers,

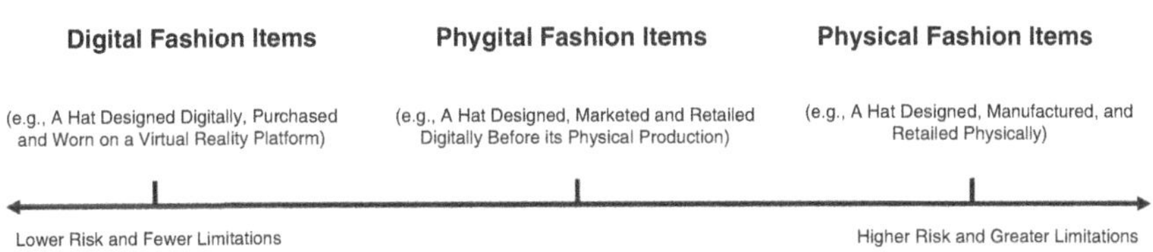

FIGURE 18.3 Proposed risk and limitations for producers and consumers of digital, phygital and physical fashion items.

(Source: author's adaptation of participant 5's framework)

as in the case of one provocateur who employs phygital techniques to decrease risk when producing new goods. This provocateur's company tests the design and market potential of a product digitally prior to its physical creation, thus reducing waste created in the sampling and marketing of a physical garment.

The differentiation between Web3 and digital fashion was another point of emphasis, with one participant stating that "Web3 has completely railroaded Digital Fashion. Digital Fashion has been defined by the Web3 aesthetic and Web3 is nothing to do with Metaverse and digital fashion" (Session 3, Participant 1). They clarified that much media attention tying the markets together resulted from investment sources. This, they argued, limits the potential of the sector due to its shaping of both industry and consumer perceptions about the purpose and benefits of digital fashion technologies.

Discussion then shifted to utility, addressing the consumer needs digital fashion can meet, for example increasing sustainability by decreasing the number of garments consumers purchase per year. Within this theme, reference was made to self-expression, online production, sampling, personalization and the ability of digital fashion to meet entertainment needs which are presently met by physical clothing. Reference was also made to tokenization, community development and the capacity to explore digital garments in minute detail. Within the varied discussions on utility, few direct references were made to sustainability. While this may have resulted from the small size of the participant sample, or bias within the group, it was an unexpected finding of the event.

Despite discussion of effective and successful use cases with positive economic and global benefits, a recurring query was "So what?", initially introduced during a discussion of the Metaverse. The participant asked,

> If you go in the Metaverse you can walk around and you can see things that look like a counterfeit future shop or something, but so what? That's the problem. I think it's a product in many cases in search of an audience.
>
> (Session 3, Participant 6)

"So what?" was reiterated throughout day with five direct references from other participants. Related discussions referred to the music, TV and film industries, emphasizing the contrasting necessity to engage with technological developments to continue listening to and watching new products. As one participant indicated,

> it's kind of easy to move with the times because as the hardware develops . . . so does the market, that's just kind of like it, given it's inbuilt. Fashion on the other hand, has always been physical. It's been on our bodies or been in that kind of . . . physical world. So, there's still that disconnect between the industry and the technology.
>
> (Session 1, Participant 1)

A further participant seconded this link:

> I think it's something about the tech-driven nature because most products in history are motivated by some kind of demand. . . . This almost feels like the opposite, in a way. It's like we can do something, but now we have to go convince people that they should buy it and that it's useful.
>
> (Session 3, Participant 7)

18.5 THE DIGITAL FASHION ECOSYSTEM

The second theme emphasized the structure of the digital fashion ecosystem. In response to a query about the composition of the field, one respondent highlighted that they believe it is still in a testing phase: "As an industry, it's still finding its feet. . . . That's why when you asked, is there a structure? I don't think there is" (Session 1, Participant 1). They indicated that much of the key focus currently remains on marketing undertaken by larger brands and international conglomerates within well-recognized platforms, supported by marketing budgets that can be targeted towards experimentation and creativity. Amidst a discussion of avatars and representation, one speaker referenced the impact of digital scans and the digital body on the modelling field. She clarified that the scanning of models within the ecommerce sector enabled businesses to cut costs and "shoot more product" (Session 2, Participant 2). This underscored that while digital avatars and scans of models' bodies can decrease the necessity to travel, cutting carbon footprints and driving forward innovation, they seem to be driving greater sales and consumption, speeding up the broader industry rather than providing opportunities to increase sustainability.

Discussions of industry actors predominantly addressed digital fashion end-products, emphasizing the important role of collectives, some with broader fashion industry support, as in the case of SKYK and its British Fashion Council (BFC) backing (Lee 2023). The role of these platforms in supporting Intellectual Property (IP) creators was highlighted as reputation building, and monetization can be challenging. The success of a few digital fashion designers was highlighted, focusing on their progression and capacity to build audiences. One participant emphasized that in this context she defines value as return on investment (ROI) and therefore focuses on specific "platforms where you can make money, and those are the gaming marketplaces – the asset source". She views asset creation for gaming platforms as "one of the biggest monetization opportunities currently", referencing democratization and the potential of trading economies for small indie creators (Session 1, Participant 3). Both gaming and Metaverse platforms were key topics throughout the course of the day. However, platform limitations were highlighted, with personal experiences including descriptions such as "really clunky" (Session 2, Participant 4).

18.6 BARRIERS TO GROWTH

A third theme that arose was barriers preventing the growth of the digital fashion field, with each relating to a requirement of digital fashion production, use or engagement. The discussion of IP creators emphasized that although a number of people have the skills to develop digital assets, the capacity to monetize remains limited, even in the case of larger digital fashion houses. This results in much of the funding and focus on digital fashion being directed towards marketing. A further challenge for IP creators related to the "Fashion 4.0" concept of interoperability – the capacity to "start having 3D assets that will work in different places", across a range of platforms and formats (Bertola and Teunissen 2018; Session 1, Participant 9). Several participants were keen on the development of technologies supporting this, which would decrease workloads within digital production, reduce emissions created within the asset creation process and provide greater purchase incentives for end users who could employ designs across varying platforms. However, debate arose in the discussion of incentives, with another participant highlighting technological, legal and financial reasons why interoperability of digital assets is challenging and unlikely to progress. The main reasoning cited for this was the lack of motivation for platforms who benefit from the lock-in effects created by present systems. While this is slowly shifting, universal interoperability has been recognized as a lasting problem within the field (VIPE Virtual Persona 2023; Tattersal 2024).

Strong interest but also hesitancy to broader fashion and retail industry adoption of digital fashion was related to technological limitations and resource requirements. Digital fashion tools require both physical and knowledge resources which may disrupt current business practices and require staff training. One keynote speaker mentioned that brands and larger fashion firms may not acknowledge the positive impact that digital tools and assets could have, specifying that it is the role of educators and digital fashion practitioners to communicate this impact to the broader industry. This speaker showcased his interpretation of the fashion industry's digital decision-making processes on an impact and resources matrix (Figure 18.4). He indicated that, in general, digital fashion is seen as either

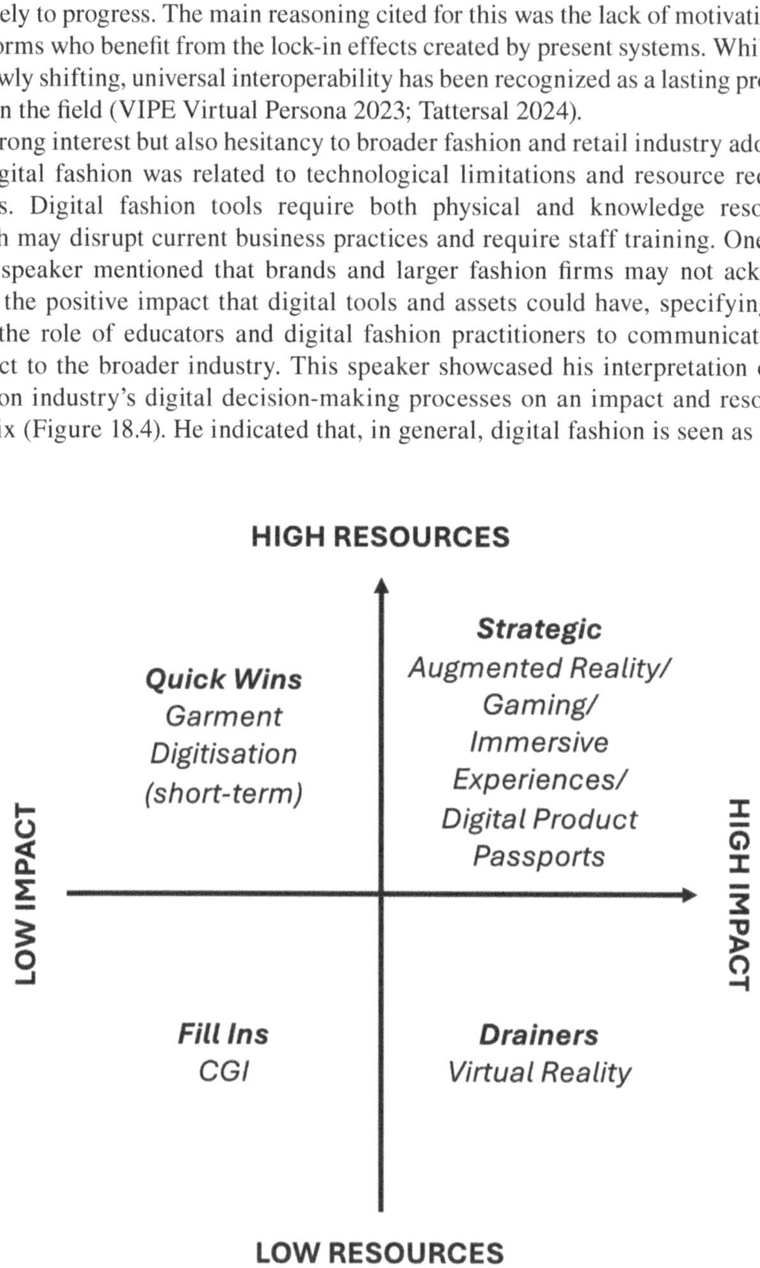

FIGURE 18.4 March 2024 State of the Industry: Brand Digital Decision-Making Challenges and Opportunities Matrix developed from Keynote Speaker 3's presentation.

(Source: author's adaptation of Keynote Speaker 3's framework)

a High Impact, High Resource "Strategic" tool or a Low Impact, High Resource "Drainer", neither of which have immediate appeal for business managers, especially in periods of uncertainty (Keynote Speaker 3). The shift to move these perceptions and increase usage of digital fashion, he indicated, relates to information and education.

Cases were presented of businesses that have employed digital fashion tools. One firm acquired an agency to digitize their garments and create digital showrooms, while another firm outsourced the work to external partners, a move related to their hesitancy to "put [in] the effort to upskill everyone" (Keynote Speaker 3). This hesitancy is common for technological innovation within fashion, and was placed within the context of growing instability within the fashion market whose larger firms are increasingly worried about their market share and resources. According to the keynote speaker, "they need to keep like the core stuff and make sure that their next quarter is good enough for their sales. So, it's, if there was something to give up when things go wrong, it's the digital and 3D side." (Keynote Speaker 3). This indicates a short-termism within industry thinking which poses a threat to digital fashion adoption.

A sense of the future potential of technological developments to aid design, innovation and increase sustainability was expressed by multiple participants. This was extended by a provocateur who posed a query about training. She indicated that fashion education predominantly focuses either on digital fashion or tradition physical skills, educating students upon a single track and discounting the potential crossover which supports phygital work. This, she underlined, poses a challenge for their recruitment of talent and limits the potential of phygital solutions. Given the sustainability and engagement benefits of phygital production, this appears to be a key barrier to future developments. Overall discussions of the role of the education sector emphasized a need to provide a broader context that allows students to understand the uses, applications and benefits of technologies and provides skills to foster lifelong learning rather than focusing on one subject area that could be learnt through open-access education.

A key discussion within the barriers theme revolved around purpose and environmental impact, with one participant highlighting the unstable situation of the fashion industry and querying whether digital fashion was " 'maybe an alternative solution to, you know, overconsumption?" She continued, asking whether digital fashion is about "marketing and generating more income", stating "if that's the case, then we have a massive issue" (Session 1, Participant 2). This discussion sparked a debate about the purpose and benefit of marketing, touching on the environmental impact and role that fashion plays in overconsumption. Another participant emphasized the important role that tokenization, an aspect he placed within the digital fashion umbrella, plays by providing details about provenance and ethical and sustainable supply chains.

This theme was followed by a provocateur who focused on the benefits offered by digital tools. She indicated that digital solutions could reduce emissions and carbon footprints, but also addressed the role played by the energy intensive infrastructure supporting them. She queried how we can employ digital solutions to reduce the industry's environmental footprint. Two further provocateurs extended this. While one highlighted production, supply chains and the development of technological hardware for consumer engagement and the other focused on storytelling, both emphasized the growing popularity and demand for digital fashion. They acknowledged the potential additional revenue stream provided by digital fashion platforms and opportunities, but clarified the importance of considering the problems that digital fashion can solve and the why behind its usage. As one indicated, it is vital to consider "our intentions when it comes to digital fashion" and "whether we use these tools to have a positive impact, help people express and enjoy themselves in the virtual world, and be more sustainable, or whether we use them to speed up production further or encourage overconsumption" (Provocation 2, Provocateur 2).

Discussions citing consumer barriers to engagement indicated a lack of knowledge, with people "not really even knowing" that they were interacting with digital fashion or what it could offer (Session 3, Participant 1). One participant suggested that the name "digital fashion" may even be "driving people" away (Session 3, Participant 9). Another participant shared findings from her research which indicated a reluctance of consumers to purchase digital fashion as the benefits were not clear to them. She suggested that "they'd rather invest in . . . something physical", clarifying that as a result of the recent cost-of-living crisis,

> it becomes a bit more of a macro issue where clothing is the last thing that they want to tangibly own. So digital fashion . . ., it would have to be free, and it would have to be for experimenting.
>
> (Session 2, Participant 2)

Further barriers emphasized the hardware required to engage with digital fashion as well as the time investments needed to successfully employ platforms. Questions of how participants engage with the digital fashion experiences they are exposed to also arose, focusing on what opportunities they have to take action and connect in a meaningful way.

18.7 KEY FINDINGS

While small in scope, and therefore difficult to generalize, this study presents findings about the nature of the digital fashion field. It clarifies the continued importance of definitions of digital fashion and how the ambiguity of language may negatively impact the sector. The collected data highlights that there are many useful and innovative forms of and applications for digital fashion, which can drive innovation, engagement and sustainability (Clean Cut TV 2024). However, it appears to stakeholders that digital fashion's boundaries, purpose and utility could benefit from greater consideration. Furthermore, the key role of differing power dynamics and interests within the digital fashion ecosystem may pose challenges, especially between platforms, larger traditional fashion companies who may or may not engage with digital fashion and individual digital fashion asset makers. Comparison can be made to the earlier platformization and resulting structure shifts within music, film and television.

Each of the identified barriers to the digital fashion field's growth are tied to a requirement of digital fashion production, use or engagement, highlighting monetization limits within the digital fashion field, despite its purported democratization of both fashion production and consumption. A key area of focus was the hesitancy of fashion brands to engage with digital fashion, illustrating an unsureness about the vital role that digital tools can play in driving forwards innovation and sustainability practices. The predominantly ancillary nature of digital fashion products was also addressed, accounting for the success of fashion within games. While opinions of future potential were voiced, emphasis was placed on the need to clarify intentions when engaging with digital fashion tools as these can either drive forward sustainability and equity or intensify overconsumption, inequality and energy usage. Participants highlighted the role of education and open information in clarifying digital fashion's benefits for industry and consumers which, in the case of industry, are often restrictively measured in ROI rather than impact on a broader or long-term scale.

18.8 CONCLUSION AND LIMITATIONS

What was apparent from this research is that there is still a lack of clarity around digital fashion and a consensus that the field is in its adolescence. Valuable use cases have been identified by both academia and practitioners from those focused on sustainability, for example, reducing textile waste within product creation, to those engaging consumers through visual representation. However, players across the sector with varying resources and power exhibit different levels of knowledge and goals, at times posing challenges to each other and the forwarding of positive innovation. Despite these, there is much opportunity for increased relevance which could not only be fostered by those involved directly in digital fashion, but also by an increased focus and potential shifts in formal and informal education, alongside engagement from interacting sectors. The present state of digital fashion is reminiscent of the music, TV, and film industries during their earlier shifts towards platform reliance, presenting similar challenges to engagement and distribution.

Limitations within this study suggest the potential of further research into the digital fashion field. This study is limited by its design as the workshop was intended to provide an initial snapshot of the field rather than in-depth analysis. The study is limited by its low number of participants and their purposive sampling. The blended nature of the event which brought together online keynotes, digital recordings and in-person sessions may also have restricted debate and discussion.

Future studies exploring the state of the digital fashion field could expand their research designs to include a broader range of participants and select a physical-only format which would better allow interaction, ideation and collective meaning development (Bryman 2008). They could also employ a longitudinal approach to account for the dramatic speed at which the industry shifts.

While this study employed a structure following the digital fashion value chain, future research could also benefit from a purely inductive approach which could better support the unexpected opinions and experiences of participants or could focus on a specific theme, such as sustainability. This work explores digital fashion predominantly as an end-product. Further work could address a broader swathe of the digital fashion sector better accounting for end-user and industry-focused digital fashion integrations. Research could consider structural and stakeholder mapping comparisons with similarly platformized sectors.

18.9 ACKNOWLEDGEMENTS

This research was supported by funding from the University of Southampton Faculty of Arts and Humanities Strategic Research Fund Round 3, 2023/2024.

REFERENCES

Arribas, Veronica, and José A. Alfaro. 2018. "3D Technology in Fashion: From Concept to Consumer." *Journal of Fashion Marketing and Management: An International Journal* 22 (2): 240–251.

Ayada, Wesam M., and Nessreen Elmelegy. 2014. "Advergames on Facebook a New Approach to Improve the Fashion Marketing." *International Design Journal* 2 (2): 139–151.

Baek, Eunsoo, Shelley Haines, Omar Fares, Zhihong Huang, Yuwei Hong, and Seung Hwan Mark Lee. 2022. "Defining digital fashion: Reshaping the field via a systematic review." *Computers in Human Behavior* 137. https://doi.org/10.1016/j.chb.2022.107407.

Bain & Company. 2023. *The Metaverse Could Reach Up to $900 Billion by 2030 But will Take Time to Scale, Finds Bain & Company*. Bain & Company. https://www.bain.com/about/media-center/press-releases/2023/the-metaverse-could-reach-up-to-$900-billion-by-2030-but-will-take-time-to-scale-finds-bain-company/.

Bendoni, Wendy K. 2017. *Social Media for Fashion Marketing: Storytelling in a Digital World*. Bloomsbury.

Bernat, Sara E. 2024. "Is Digital Fashion Socially Equitable?" In *Digital Fashion: Theory, Implications, Practice*, edited by Michael Spicher, Sara E. Bernat, and Doris Domoszlai-Lantner. Bloomsbury.

Bertola, Paola, and Jose Teunissen. 2018. "Fashion 4.0. Innovating Fashion Industry through Digital Transformation." *Research Journal of Textile and Apparel* 22 (4): 352–369.

Black, Sandy. 2018. "Sustainability and Digitalization." In *The End of Fashion: Clothing and Dress in the Age of Globalization*, edited by Adam Geczy, and Vicki Karaminas. Bloomsbury Visual Arts.

Boughlala, Adil, and Anneke Smelik. 2024. "Tracing the History of Digital Fashion." *Clothing and Textiles Research Journal*. https://doi.org/10.1177/0887302X241283504

Bryman, Alan. 2008. *Social Research Methods*. Oxford University Press.

Business of Fashion, and McKinsey & Company. 2023. *The State of Fashion 2024: Riding Out the Storm*. McKinsey & Company. https://www.businessoffashion.com/reports/news-analysis/the-state-of-fashion-2024-report-bof-mckinsey/.

Byrne, D. 2021. "A Worked Example of Braun and Clarke's Approach to Reflexive Thematic Analysis." *Quality & Quantity* 56. https://doi.org/10.1007/s11135-021-01182-y.

Chan, Hazel Hoi Yau, Claudia Henninger, Rosy Boardman, and Marta Blazquez Cano. 2023. "The Adoption of Digital Fashion as an End Product: A Systematic Literature Review of Research Foci and Future Research Agenda." *Journal of Global Fashion Marketing* 15 (1): 155–180. https://doi.org/10.1080/20932685.2023.2251033.

Citi GPS. 2022. *Metaverse and Money: Decrypting the Future*. Citi Global Perspectives and Solutions. https://www.citigroup.com/global/insights/citigps/metaverse-and-money_20220330.

Clean Cut TV. 2024. *Fash-Forward Podcast*. "Fixing the Fashion Industry Through AI and Social Innovation." Posted March 25, 2024 by Clean Cut TV. YouTube, 44:30. https://www.youtube.com/watch?v=fQCPVGzL8yY.

Colombi, Chiara, and Erminia D'Itria. 2023. "Fashion Digital Transformation: Innovating Business Models towards Circular Economy and Sustainability." *Sustainability* 15 (6). https://doi.org/10.3390/su15064942.

Digital Fashion Designer's Council. 2024. "Homepage." https://www.dfdc.xyz.

Digital Fashion Week. 2024. "Digital Fashion Week." https://digitalfashionweek.nyc.

Drinkwater, Mathew. 2024. "Mathew Drinkwater Interviewed by Sara Emilia Bernat." In *Digital Fashion: Theory, Implications, Practice*, edited by Michael Spicher, Sara E. Bernat, and Doris Domoszlai-Lantner. Bloomsbury.

Farren, Anne, and Andrew Hutchison. 2004. "Cyborgs, New Technology, and the Body: The Changing Nature of Garments." *Fashion Theory* 8 (4): 461–475. https://doi.org/10.2752/136270404778051618.

Future+ and Luxurynsight. 2023. *New Codes*. Future+. https://www.futureplus.xyz/media/new-codes-report.

Future+ and Luxurynsight. 2024. *Codes of Culture Insights Report*. Future+. https://www.futureplus.xyz/media/report-coc.

Hanson, Ben. 2023. "MVFW 2023: Managing the Fallout". *The Interline*, May 9. https://www.theinterline.com/2023/05/09/metaverse-fashion-week-2023-managing-the-fallout/#:~:text=Compared%20to%20the%20same%20event,the%20event%2C%20was%20around%201%2C000.

Harris, Jane. 2005. "Crafting' Computer Graphics – a Convergence of Traditional and 'New' Media." *Textile* 3 (1): 20–35. https://doi.org/10.2752/147597505778052666.

Heim, Hilde. 2022. "Digital Fashion Revolutions: Supply Chain Transparency, Digitalization and the Non-Disclosure Paradox." *Fashion Practice* 14 (3): 329–351. https://doi.org/10.1080/17569370.2022.2118975.

Ihenacho, Emeka. 2024. "Impact of Digital Fashion Shows on Consumer Engagement with Fashion Brands in Nigeria." *International Journal of Fashion and Design* 3 (3): 25–34. https://doi.org/10.47604/ijfd.2862.

Janssens, Alice, and Mariangela Lavanga. 2018. "An Expensive, Confusing, and Ineffective Suit of Armor: Investigating Risks of Design Piracy and Perceptions of the Design Rights Available to Emerging Fashion Designers in the Digital Age." *Fashion Theory* 24 (2): 229–260. https://doi.org/10.1080/1362704X.2018.1515159.

Keh, Pei-Ru. 2023. "Ralph Lauren Launches the 888 House, a Surreal Digital Store Fusing Fantasy and Reality." *Wallpaper*, August 22. https://www.wallpaper.com/fashion-beauty/ralph-lauren-the-888-house-digital-store.

Lee, Adriana. 2023. "EXCLUSIVE: SYKY Partners with British Fashion Council on Collective Incubator." *WWD*, June 27. https://wwd.com/business-news/technology/exclusive-syky-collective-british-fashion-council-incubator-1235715089/.

Loftus, Dani. 2024a. "#011 Digital Fashion in 2023: An Obituary to a False Start." *This Outfit Does Not Exist*. Substack, January 11, 2024. https://thisoutfitdoesnotexist.substack.com/p/digital-fashion-in-2023-an-obituary.

Loftus, Dani. 2024b. "#012 The Year for Digital Fashion is 2024." *This Outfit Does Not Exist*. Substack, April 1, 2024. https://thisoutfitdoesnotexist.substack.com/p/the-year-for-digital-fashion-is-2024.

Luce, Leanne. 2018. *Artificial Intelligence for Fashion: How AI Is Revolutionizing the Fashion Industry*. Apress.

Makryniotis, Thomas. 2018. "Fashion and Costume Design in Electronic Entertainment – Bridging the Gap between Character and Fashion Design." *Fashion Practice* 10 (1): 99–118. https://doi.org/10.1080/17569370.2017.1412595.

Mattila, Heikke. 2016. "Digital Fashion – How and When?" *Tekstil: Journal of Clothing and Textile Technology* 65 (9–10): 334–339.

McGimpsey, Patrick, and Johanna Leggatt. 2024. "The Rise and Fall of NFTS: What Went Wrong?" *Forbes*, January 31. https://www.forbes.com/advisor/au/investing/cryptocurrency/are-nfts-dead/.

Monahan, Cait. 2024. "Digital Fashion Collaborations You Need to Know." *Hypebae*, January 2. https://hypebae.com/2023/12/digital-fashion-campaigns-collaborations-paolina-russo-roblox-rtfkt-ledger.

Noris, Alice, Tekila Harley Nobile, Nadzeya Kalbaska, and Lorenzo Cantoni. 2020. "Digital Fashion: A Systematic Literature Review. A Perspective on Marketing and Communication." *Journal of Global Fashion Marketing* 12 (1): 32–46. https://doi.org/10.1080/20932685.2020.1835522.

Park, Juha, and Jaehoon Chun. 2023. "Evolution of Fashion as Play in the Digital Space." *Fashion Practice* 15 (2): 256–278. https://doi.org/10.1080/17569370.2022.2149837.

Periyasami, Aravin Prince, and Saravanan Periyasami. 2023. "Critical Review on Sustainability in Denim: A Step toward Sustainable Production and Consumption of Denim." *ACS Omega* 8 (5): 4472–4490.

Rocamora, Agnès. 2016. "Mediatization and Digital Media in the Field of Fashion." *Fashion Theory* 21 (5): 505–522. https://doi.org/10.1080/1362704X.2016.1173349.

Särmäkari, Natalia. 2021a. "Digital 3D Fashion Designers: Cases of Atacac and The Fabricant." *Fashion Theory* 27 (1): 85–114. https://doi.org/10.1080/1362704X.2021.1981657.

Särmäkari, Natalia. 2021b. "'Digital Fashion' on Its Way from Niche to the New Norm." In *'The New Normal': Sartorial and Body Practices of the Quarantine Era ["НОВАЯ НОРМА": гардеробные и телесные практики в эпоху пандемии]*, edited by L. Alibieva. The New Literary Observer, Fashion Theory Russia.

Saxena, Aarushi. 2022. "Digital Fashion Claims to Solve Overconsumption and Overproduction: How Green is It Really?" *Lampoon*, February 22. https://lampoonmagazine.com/article/2022/02/22/digital-fashion/.

Sayem, Abu Sadat Muhammad, Samit Chakraborty, S. M. Azizul Hoque, Kowshik Saha, Mushfika Tasnim Mica, and Mominul Ahsan. 2023. "Defining Digital Fashion and Tracking the Development of Relevant Technologies." In *Digital Fashion Innovations: Advances in Design, Simulation and Industry*, edited by Abu Sadat Muhammad Sayem. CRC Press. https://doi.org/10.1201/9781003264958.

Sin Ying, Ng, and P. Y. Mok. 2023. "A Review of Blockchain Technology for Sustainable Fashion." In *Digital Fashion Innovations: Advances in Design, Simulation and Industry*, edited by Abu Sadat Muhammad Sayem. CRC Press. https://doi.org/10.1201/9781003264958.

Skjulstad, Synne. 2018. "Vetements, Memes, and Connectivity: Fashion Media in the Era of Instagram." *Fashion Theory* 24 (2): 181–209. https://doi.org/10.1080/1362704X.2018.1491191.

Spicher, Michael, Sara Emilia Bernat, and Doris Domoszlai-Lantner. 2024. "An Introduction to Digital Fashion." In *Digital Fashion: Theory, Implications, Practice*, edited by Michael Spicher, Sara E. Bernat, and Doris Domoszlai-Lantner. Bloomsbury.

Tattersal, Clare. 2024. "Games X Fashion > Unpacking the Audience and Analysing the Value Proposition." *The Interline*, April 3. https://www.theinterline.com/2024/04/03/games-x-fashion-unpacking-the-audience-and-analysing-the-value-proposition/.

Van Volkinburg, Nina. 2020. "Disruption to Destruction: Exploring the Effects of Digital Disruption on the Value Creation Processes within the Field of Fashion through the lens of Service Dominant Logic." PhD diss., University of Exeter. https://ore.exeter.ac.uk/repository/handle/10871/123467.

VIPE – Virtual Persona. 2023. "VRM Avatar File: Enabling Interoperability across the Metaverse." *Medium*, June 21. https://medium.com/@vipeio/vrm-avatar-file-enabling-interoperability-across-the-metaverse-359091346cbd.

Volino, Pascal, Frederic Cordier, and Nadia Magnenat-Thalmann. 2005. "From Early Virtual Garment Simulation to Interactive Fashion Design." *Computer-Aided Design* 37 (6): 593–608. https://doi.org/10.1016/j.cad.2004.09.003.

von Wachenfeldt, Paula 2019. "The Mediation of Luxury Brands in Digital Storytelling." *Fashion Theory* 25 (1): 99–118.https://doi.org/10.1080/1362704X.2019.1599256.

Wilczek, Beata. 2024. "(Im)possible Digital Fashion Dreams: Who Can Deliver on Accessible and Sustainable Digital Fashion?." In *Digital Fashion: Theory, Implications, Practice*, edited by Michael Spicher, Sara E. Bernat, and Doris Domoszlai-Lantner. Bloomsbury.

Zhang, Yanbo, Chuanlan Liu, and Yanru Lyu. 2023. "Examining Consumers' Perceptions of and Attitudes towards Digital Fashion in General and Purchase Intention of Luxury Brands' Digital Fashion Specifically." *Journal of Theoretical and Applied Electronic Commerce Research* 18 (4): 1971–1989. https://doi.org/10.3390/jtaer18040099.

Zwieglinska, Zofia. 2024. "As London Fashion Week Celebrates 40 Years, the BFC Turns to Supporting Phygital Designers." *Glossy*, February 16. https://www.glossy.co/fashion/as-london-fashion-week-celebrates-40-years-the-bfc-turns-to-supporting-phygital-designers/.

Index

A

B

C

D

E

F

G

H

T

U

V

W

Y

Z